Adolf Alt

A Treatise on Ophthalmology for the General Practitioner

Adolf Alt

A Treatise on Ophthalmology for the General Practitioner

ISBN/EAN: 9783337800222

Printed in Europe, USA, Canada, Australia, Japan

Cover: Foto ©berggeist007 / pixelio.de

More available books at **www.hansebooks.com**

A TREATISE

ON

OPHTHALMOLOGY.

FOR THE

GENERAL PRACTITIONER.

SECOND EDITION, REVISED AND ENLARGED.

WITH 140 ILLUSTRATIONS.

BY

ADOLF ALT, M. D.

J. H. CHAMBERS & CO.,
PUBLISHERS AND DEALERS IN MEDICAL BOOKS, ST. LOUIS.

PREFACE.

The manner in which the first edition has been received by the public, for which it was intended, rendered a new edition of this book necessary several years ago. I have only now been able to rewrite it.

Although this second edition is considerably enlarged, I have tried to adhere closely to the original plan, to make it a useful guide and a help to the *general practitioner* who may be forced to take care of certain eye diseases. That this plan has proven a very acceptable one, I have received ample evidence.

In this second edition I have eliminated the illustrations of eye-instruments, as every instrument-makers' catalogue contains them. On the other hand I have tried to give particularly a larger number of practical illustrations. Those of them, which are not my own, have, as far I was able to find it, been credited to their proper source. A few of them have been copied and recopied so often in different books, that their original source could not be determined.

Like the first edition, this second one is not intended for the specialist. That it may be useful to the general practitioner, has been and is my desire.

June, 1893.   ADOLF ALT.

CONTENTS.

CHAPTER I.

ANATOMY OF THE EYE.

CHAPTER II.

EXAMINATION OF THE EYE.

CHAPTER III.

DISEASES OF THE EYELIDS.

CHAPTER IV.

DISEASES OF THE LACHRYMAL APPARATUS.

CHAPTER V.

DISEASES OF THE ORBIT.

CHAPTER VI.

MINOR MANIPULATIONS IN THE TREATMENT OF EYE-DISEASES.

CHAPTER VII.

DISEASES OF THE ORBIT.

CHAPTER VIII.

DISEASES OF THE CORNEA.

CHAPTER IX.

DISEASES OF THE SCLEROTIC.

CHAPTER X.

DISEASES OF THE IRIS.

CHAPTER XI.

DISEASES OF THE CILIARY BODY.

CHAPTER XII.

DISEASES OF THE CHOROID.

CHAPTER XIII.

DISEASES OF THE RETINA.

CHAPTER XIV.

DISEASES OF THE OPTIC NERVE.

CHAPTER XV.

DISEASES OF THE CRYSTALLINE LENS.

CHAPTER XVI.

DISEASES OF THE VITREOUS BODY.

CHAPTER XVII.

GLAUCOMA.

CHAPTER XVIII.

INJURIES OF THE EYEBALL AND THEIR CONSEQUENCES.

CHAPTER XIX.

SYMPATHETIC OPHTHALMIA.

CHAPTER XX.

ERRORS OF REFRACTION AND ACCOMMODATION.

CHAPTER XXI.

DISEASES OF THE EXTERNAL MUSCLES OF THE EYE.

CHAPTER XXII.

ON THE DIAGNOSTIC VALUE OF EYE-DISEASES IN INTRA-CRANIAL AFFECTIONS.

CHAPTER XXIII.

DEVELOPMENT OF THE EYE AND CONGENITAL MALFORMATIONS.

ILLUSTRATIONS.

OPHTHALMOLOGY.

CHAPTER I.—ANATOMY OF THE ORBIT, EYE, AND ITS ADNEXA.

§1. *The orbit* is a quadrilateral pyramidal cavity surrounded by bony walls, which separate it upwards from the cranial and frontal cavities, downwards from the antrum Highmorii and inwards from the nasal cavity. (See Fig. 1).

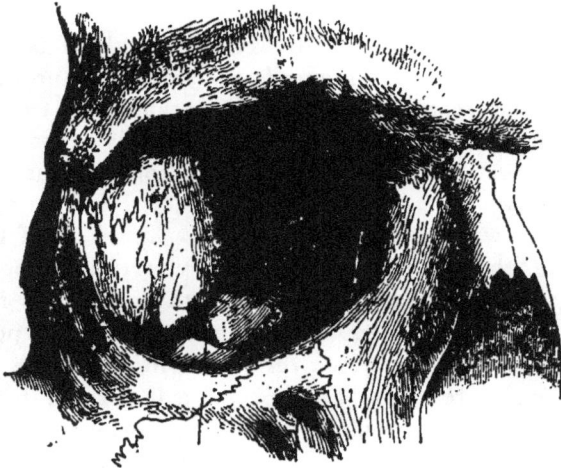

FIG. 1.—(After Merkel). Right orbit viewed from in front. Showing the upper orbital fissure to the left and upward and next to it, on the right, the optic canal. To the left and downward is seen the lower orbital fissure.

The base of this pyramid is formed by the large opening in front. Its apex lies in the depth, and is also represented by several openings.

The rim around the large opening in front is called the orbital margin, and we speak of an inner, upper, outer and lower orbital margin. The last three of these margins are slightly overhanging the interior of the orbit and have a sharp edge, whilst the inner one is rounded off. The outer, upper and lower margins are formed of very hard and dense bone.

Seven different bones of the skull and face help together in making up the walls of the orbit. The upper wall is formed by the horizontal portion of the frontal bone and a small part of the small wing of the sphenoidal bone. On its temporal side, behind the superior orbital margin we find in this wall a small depression in which the lachrymal gland is situated, which is called the *fossa glandulæ lacrymalis*. Another small depression lies near the median side, in which the trochlea for the superior oblique muscle is found. This is called the *fossa trochlearis*. Above this wall lies the frontal sinus and the cranial cavity.

The inner wall consists of the *lamina papyracea* of the ethmoid bone, the lachrymal bone, and a very small portion of the body of the sphenoid bone near the apex. In its most anterior part it has a deep depression with a sharp posterior and anterior edge, the *fossa lacrymalis* in which the lachrymal sack is situated. Its anterior edge is formed by a projection of the supramaxillary bone, the posterior one by the lachrymal bone. These two edges are joined farther down and form the lachrymal canal for the lachrymal duct. In this orbital wall we find two openings, the foramen ethmoidale anterius for the nervous ethmoidalis and the anterior vasa ethmoidalia, and the foramen ethmoidale posterius for the posterior vessels of the same name.

The lower wall which is higher on its nasal than on its temporal side, consists of the orbital portion of the supramaxillary bone, the orbital portion of the zygomatic bone, and the orbital portion of the palatine bone. It has a longitudinal groove near its middle line, *sulcus infraorbitalis*, in which the infraorbital nerve and artery pass forward into the infraorbital canal. Below this wall lies the *antrum* of Highmore.

The outer wall consists of the temporal portion of the zygo-

matic bone and of the temporal wing of the sphenoidal bone.
In it we find a groove for the nervus zygomatico-temporalis
and one or two openings for the nèrvus zygomatico-facialis.

The upper orbital margin has near its nasal end a large incision for the supraorbital nerve and vessels, and a smaller one
for the frontal nerve which, coming out of the orbit, go to
the forehead.

Below the nasal end of the lower orbital margin is the opening of the infraorbital canal through which the infraorbital
nerve and artery come to the surface.

At the apex of the orbital pyrámid lies the *optic foramen*
or canal, a short funnel-shaped passage for the optic nerve and
ophthalmic artery. The direction of this canal is up—and
inward.

There are, furthermore, near the apex and converging towards it two fissures where the outer wall is joined to the lower
and to the upper one. They are called the upper and lower
orbital fissures, or the *sphenoid* and *spheno-maxillary fissures.* Through the upper fissure pass the motor nerves of
the eye, and the ophthalmic nerve and vein. On its inner
side within the skull lies the cavernous sinus. These fissures
vary in height and breadth.

The inner walls of the two orbits run nearly parallel to each
other, and in consequence the axes of the two cavities converge towards the median line.

The *periosteum* of the orbital cavity is formed by the dura
mater.

This membrane, after having entered the orbit through the
canalis opticus and fissura sphenoidea, is split into two parts,
one of which serves to form the periosteal coat of the orbit,
while the other, in the main, forms the dura mater sheath of
the optic nerve, and a capsule for the posterior parts of the
eye-ball, called *Tenon's capsule.* This capsule ensheathes about
the posterior four-fifths of the eye-ball leaving a small space bare
where the optic nerve and the ciliary nerves and blood-
vessels enter it, and ends in front near the limbus conjunctivæ
by joining this membrane. It is a serous membrane, lined
with a layer of endothelial cells. The serous space, which

lies between it and the eye-ball, is called *Tenon's space.* (See Fig. 2).

In this capsule the eye moves very much like a joint in its capsule, although numerous fibres traverse this space, and insert themselves in the episcleral tissue and Tenon's capsule, and *vice versa.*

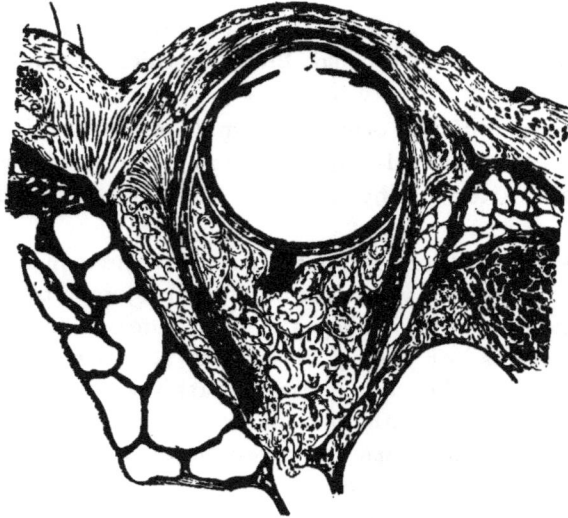

FIG. 2.—(After Gerlach). Horizontal section through lids, eyeball and orbit, showing their relative positions, also Tenon's capsule and the lymph-sheaths for the internal and external rectus muscles.

From Tenon's capsule a large number of small trabeculæ run back into the periosteum of the orbit. Between these trabeculæ lies the orbital fat. In the neighborhood of the lachrymal gland they contribute to the formation of its firm fibrous capsule. They, furthermore, help to keep the eye-ball and the other contents of the orbit in position.

Tenon's space can be inflated or injected from the subdural space of the cranium. Such an injection shows that this space ends near the corneo-scleral margin, where the tissue of Tenon's capsule goes over into the tissue of the ocular conjunctiva.

The external muscles of the eye-ball, of which there are six (the *rectus superior, inferior, externus* [*abducens*] and *internus*, and the *obliquus superior* and *inferior*), must naturally pierce this capsule to reach their insertions in the sclerotic, and they receive a sheath from it. The sheaths of the recti muscles can be traced backward into the orbital fat where they are gradually lost in the perimysium. The sheath of the superior oblique (*trochlearis*) muscle reaches to the trochlea, and there joins the periosteum, while that of the inferior oblique muscle hardly reaches as far back as the orbital adipose tissue (*Gerlach*).

The layer of the dura mater which forms the *periosteum of the orbit,* runs forward to the anterior margins of that cavity, where it passes over into the periosteum of the surrounding bones. It also gives off a fascial layer for the eye-lids, called the *tarso-orbital fascia.* The orbital periosteum is, for the most part, only loosely connected with the bone, but wherever there is an aperture in the orbital walls, and also at the orbital margins, its attachments are very firm.

§2. The *eye-lids* are originally a duplicature of the skin, growing down from the upper and up from the lower orbital margins during foetal life. The part of this fold which lies directly upon the eye-ball takes on the character of a mucous membrane, the *palpebral conjunctiva,* and forms with the *ocular conjunctiva* a cul-de-sac, which is called the *fornix* of of the conjunctiva.

At the free margin of the lids this mucous membrane and the cutaneous outer surface pass over into each other, in the same way as they do, for instance, on the lips. The free margins of the eye-lids form two distinct edges, the inner one (toward the eye) sharp, the outer one rounded off. Where the upper and lower eye-lids join each other in the horizontal line, they form the outer and inner angles of the palpebral fissure, (*outer and inner canthus*). The outer angle is sharp; the inner is rounded. Behind the inner angle of the palpebral fissure lies a small reddish, round body, called the *lachrymal caruncle.* It has the structure of the cutis, and contains fine hairs and sebaceous glands. The fatty secretion of these glands stems the current of the lachrymal fluid and helps to direct it to the

channels of drainage. On the temporal side of the caruncle the
conjunctiva forms a semilunar fold with the concavity directed
toward the cornea which is a remnant of the third lid or, *membrana nictitans,* as we find it in animals. It is called the *plica
semilunaris.*

A little outwards from the inner angle of the palpebral fissure each eye-lid shows at the inner edge of the margin a
small papilla-like elevation with a small aperture at its apex.
These elevations are the *lachrymal papillæ;* the apertures are
the *lachrymal puncta.*

The cutis of the eye-lids is thin and its hairs are very fine
and short. The subcutaneous tissue is very loose and contains
no fat.

Between the conjunctival and cutaneous surfaces of the lids
lie the tarsal tissue, the muscular layer, nerves and blood-vessels.

The *tarsal tissue,* commonly called the tarsal cartilage, lies
close upon the conjunctiva. It consists of dense, tendon like
connective tissue and is really no cartilage. This tarsal tissue,
freed from its surroundings, has a more or less semi-lunar
shape. In the upper eye-lid its convexity is directed upwards,
in the lower eye-lid, downwards.

Near the conjunctival surface a number of glands, the *Meibomian* or *tarsal glands,* lie embedded in the tarsal tissue, in a
direction more or less at right angles to the lid margins. The
orifices of the ducts of those glands are arranged in a row, at
the inner edge of the free margin of each eye-lid. Their secretion is a fatty substance.

Nearer the outer edge of the free margin of the eye-lid
grow the *cilia, eyelashes.* They are short, strong hairs, which
are curvilinear in form, and are so directed that those of the
upper and lower eye-lids turn their convexities toward each
other, those of the upper lids being convex downward, those
of the lower lids convex upward. They differ from other hair
by the fact, that they live on an average only for a period of
of from 60 to 100 days, and then drop off.

The conjunctiva of the lids is closely attached to the tarsal
tissue, no submucous layer intervening. Where the tarsal tissue
ends, submucous tissue makes its appearance, being very loose

and of an adenoid character. This adenoid tissue is thickest in the fornix of the conjunctiva. In this region the surface of the conjunctiva is wrinkled and folded, and numerous muciparous (*Krause and Waldeyer*) glands open into it.

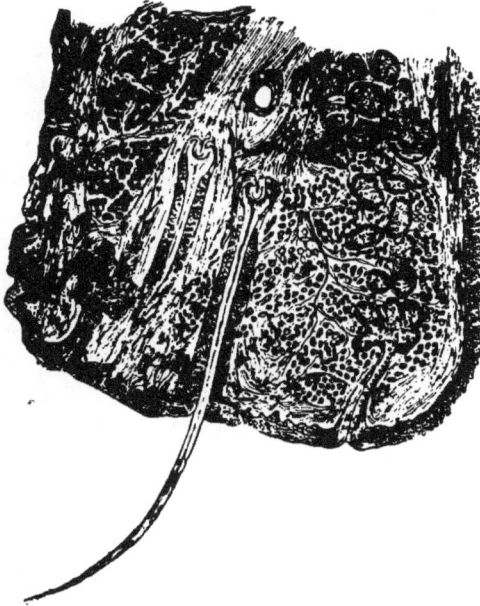

FIG. 3.—(After Gerlach).—The tissues near the margin of the upper lid. To the left, the cutaneous, to the right, the conjunctival surface. Next to the conjunctival surface lies the tarsal tissue in which is seen embedded a Meibomian gland with its efferent duct opening at the inner edge of the lid margin; around it are seen the fibres of the muscle of Riolanus. Nearer the cutaneous edge two cilia are seen and a modified sudoriferous gland. To the right of the cutaneous surface lies a portion of the orbicularis palpebrarum muscle.

The muscles of the eye-lids are embedded in the loose connective tissue on the outer surface of the tarsus. The most important one is the *orbicularis palpebrarum*. This is a very broad, thin muscle, covering the whole area of the eye-lids, and reaching somewhat beyond them in all directions. It consists of three component parts which are called the palpebral, the orbital and the malar portions. The orbicularis acts as a sphincter muscle, contracting the palpebral fissure and closing the eye-lids.   A small portion of this muscle, which lies in the tissue between the roots of the eyelashes and the excretory ducts of the Meibomian glands, is called the *ciliary muscle* of *Riolanus*.   Its function seems to be to help in moving the secretion of the Meibomian glands to the lid-margin. (See Fig. 3).

There is, furthermore, a non-striated muscle situated in both upper and lower lids.   It is very thin, but almost as wide as the lid, lies near the conjunctival fornix, and its fibres run at right angles to the lid-margin.   It ends on one hand in the tarsus, on the other in the subcutaneous tissue.   After its discoverer it is called *Mueller's* muscle, or the superior and in-inferior palpebral muscle.

Where the upper and lower halves of the orbicularis muscle join each other at the outer and inner angles of the palpebral fissure,they form the *ligamentum palpebrale externum and internum*, of which, however, only the inner one is a real ligament.

At the upper edge and along the whole breadth of the tarsal tissue of the upper eye-lid the *levator palpebræ superioris muscle* is inserted by a broad, thin tendon.   This muscle rolls the uper eye-lid upward and backward into the orbit, and thus opens the eye.

§3.   The *ocular conjunctiva* begins at the fornix and ends at the cornea-scleral margin (*limbus corneæ*).   Its submucous adenoid tissue is loosely connected with the sclerotic (*episcleral tissue*).   No glands are found in the ocular conjunctiva, although its epithelial layer contains numerous mucoid cells.

The shape of the eye-ball is nearly spherical, and is determined by the so-called hard membranes which together constitute its outer walls, namely, the sclerotic and the cornea.

§4.   The *sclerotic* consists, like the tarsus, of a dense connective tissue, the fibres of which are irregularly interwoven, and are held together by a protoplasmic cementing substance. Embedded in this latter is a system of lymphatic canals, which enlarge at intervals and contain large, flat, stellated connective-tissue cells.   It, furthermore, contains nerves and blood-vessels.

The fibres of the sclerotic run mostly in an approximately longitudinal (*meridional*) direction. Fibres running in a circular (*æquatorial*) direction are found in larger quantities only around the optic nerve entrance and near the corneo-scleral margin.

At the optic nerve entrance the sheaths of the nerve become merged in the sclerotic. There is no large opening in the latter membrane to admit the optic nerve, as a whole, into the eye-ball, but a large number of small holes, each admitting a bundle of nerve-fibres. This sieve-like region is called the *lamina cribrosa* of the sclerotic. The tendons of the external muscles of the eye-ball are lost in the tissue of the sclerotic at their insertions.

The sclerotic has an endothelial coat on its outer and inner surface, and is pierced by the ciliary nerves and arteries, and by the *venæ vorticosæ*, with their respective lymph-sheaths.

At the corneo-scleral margin the tissue of the sclerotic, which is only translucent, passes over into the transparent tissue of the cornea, but in such a manner that the sclerotic tissue slightly overlaps the cornea at its periphery. The manner of the junction of these two membranes is best likened to the way in which a watch-crystal sits in the rim of the watch.

§5. The *cornea* consists of fibres of a perfectly transparent modified connective tissue, which are regularly arranged in bundles, and these again in lamellæ, which lie more or less parallel to each other, and are all united by the same protoplasmic cementing substance, which is found in the sclerotic. In this substance are enclosed the lymphatic canals of the cornea, which, like the lamellæ, are more numerous and lie closer together toward the anterior surface of the cornea. They have, like the scleral canals, numerous ampulla-like enlargements (*lacunæ*), in which are contained the large, flat many-branched connective-tissue cells of the cornea (*corneal corpuscles*). At the corneo-sceral margin this system of canals goes directly over into the similar system of canals in the sclerotic.

Near the outer (anterior) surface the layers of the cornea become more compact, and finally coalesce to form a layer, which by its lack of cellular elements appears like a distinct hyaline, elastic membrane. This is called *Bowman's or Reich-*

ert's layer. (See Fig. 4) At its posterior surface the cornea is lined by a thin vitreous membrane called *Descemet's* membrane. This is an elastic membrane, and rolls upon itself, when divided or separated from the corneal tissue.

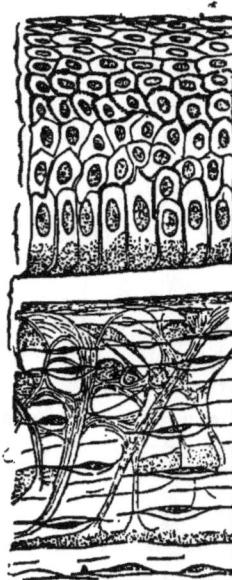

FIG. 4.—(After Waldeyer). Meridional section through the superficial layers of the cornea of the calf. Shows: Flattened epithelial cells; prickle and polymorphous cells; basal (club-shaped) cells. Below the epithelium, Bowman's layer; below this, the corneal tissue proper with canals, lacunæ, fixed and wandering cells.

Upon the anterior surface of *Bowman's* layer lies the corneal epithelium. *Descemet's* membrane on its posterior surface is lined by a single layer of endothelial cells.

At the corneo-scleral margin, where the ocular conjunctiva ends, its epithelium goes directly over into the epithelium of the cornea. *Bowman's* layer, together with the nearest corneal lamellæ, is split into fibrillæ and becomes merged in the subconjunctival tissue. *Descemet's* membrane, with the adjoining layers of the cornea, is similarly split into fibrillæ at

the periphery of the cornea, and is lost partially in the tendon
of the ciliary muscle and partially in the iris. On their way

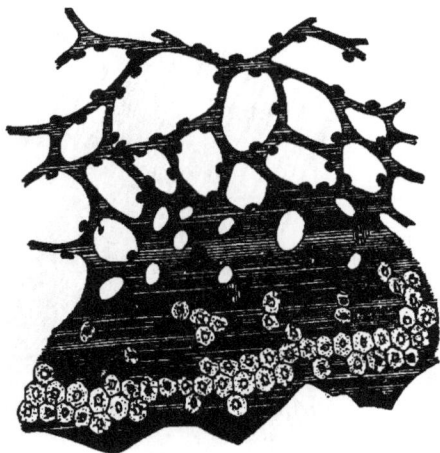

FIG. 5.—A portion of Descemet's membrane partially covered with endothelial cells
torn from the cornea. Its fibres form (upwards) the network of the so-
called ligamentum pectinatum of the iris.

these fibres form what is called the *ligamentum pectinatum* of
the iris. (See Figs. 5 and 6). Between the fibres of this so-
called ligament lie a large number of cavities, which are called
Fontana's cavities. These cavities communicate toward the
outer surface with the canalicular system of the cornea and
sclerotic, and with *Schlemm's* canal, a larger-lymph canal embedded in the corneo-scleral tissue; on the other side they
open into the anterior chamber.

The corneal tissue contains blood-vessels only at its periphery, where a system of loops of capillaries reaches into it for
the distance of about one millimeter. There seem to be two
sets of these loops, one in the deeper layers of the cornea
and one more superficially situated. The arterial vessels which
take part in the formation of these loops come from the anterior ciliary arteries. and anastomose with the blood-vessels of
the conjunctiva. The blood is carried away from these loops
by small veins which empty it into the episcleral and by this
route into the anterior ciliary veins.

The cornea at its periphery is, furthermore, supplied by a number of larger nervous branches which come from the conjunctival and anterior ciliary nerves. They enter near the pos-

FIG. 6.—Meridional section through the corneo-scleral margin and iris-angle, showing the manner in which the tissues of the cornea join those of the sclerotic, iris, and tendon of the ciliary muscle. The points of interest are: Schlemm's canal, the large lymph-space on the inner side of the sclerotic, and the fibres into which Descemet's membrane is split up, which form the ligamentum pectinatum with Fontana's spaces lying between meshes.

terior surface of the cornea and lie in a special system of canals. Soon after having entered the corneal tissue, the nerve-fibres lose their double contour, and the main stems give of branches which soon form a network, called the deep stroma-plexus. From this plexus smaller branches rise towards the surface of the cornea, split into axis-cylinders and axis-fibrillæ, and after having formed another network under Bowman's layer, called the superficial stroma-plexus, they pierce this layer nearly at a right angle, and form a third network between the epithelial cells (the intra-epithelial plexus), and are there lost. (See Fig. 7).

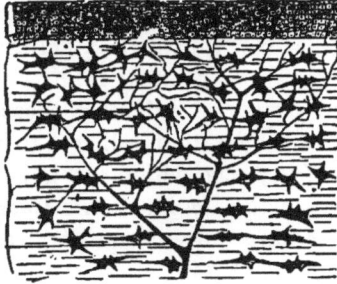

Fig. 7.—(After Waldeyer). Oblique section of the human cornea stained with chloride of gold, in order to show the distribution of the smaller nerve-branches within its tissue and epithelium.

§6. Next to the inner surface of the sclerotic lies the *uveal tract*, the vascular membrane of the eye-ball. Although the uveal tract consists, in the main, of the same tissue from one end to the other, it is divided into three distinct parts, the *choroid*, the *ciliary body* and the *iris*.

Fig. 8.—(After Merkel). The sclerotic being removed, the manner in which the ciliary nerves pass through the suprachoroidal space to reach the anterior parts of the eye, is shown.

The uveal tract firmly adheres to the sclerotic around the

optic nerve entrance and at the cornea-scleral margin. Between these two attachments it is slightly separated from the sclerotic by the supra-choroidal space. This space is traversed by innumerable fibres going from the uveal tract into the sclerotic and vice versa, which thus form a delicate, spongy tissue containing a great many endothelial cells. When the choroid is forcibly detached from the sclerotic these fibres are torn, and the part of them which then adheres to the sclerotic has been called the *lamina fusca*, while the part adhering to the choroid, is known as the *lamina suprachoroidea*. In this spongy tissue the ciliary nerves run forward to the ciliary body, after having pierced the sclerotic near the entrance of the optic nerve. (See Fig. 8).

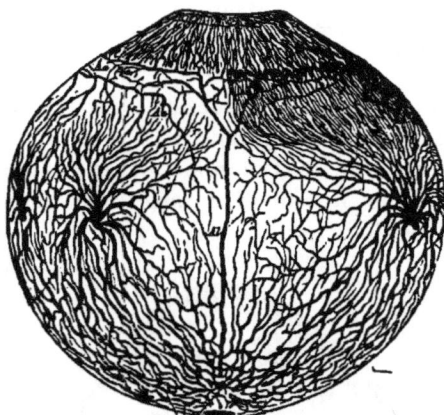

FIG. 9.—(After Leber). Distribution of bloodvessels in the uveal tract.

The *choroid* proper consists of a loose network of connective-tissue fibres, which contains a large number of stellated pigmented and unpigmented cells. The pigmented cells are more numerous in the outer two-thirds of the choroid, their pigment varying considerably in tint in different eyes. In albinos it may be wanting or it is slightly yellowish, in negroes deep brown, or even black; and all intermediate shades may be seen in different eyes, corresponding in a general way with the pigmentation of the skin and hair of the individual.

In this loose network of connective tissue lie embedded the

innumerable blood-vessels of the choroid, which come from the short posterior ciliary arteries. The veins which collect the blood and empty it into four or six larger trunks, the *venæ vorticosæ*, lie in the outer two-thirds of the choroid. The inner third contains the capillaries (*chorio-capillaris*). (See Fig. 9).

The choroid contains, moreovor, a large number of nerves and ganglionic cells, and some organic muscular fibres. On its inner surface it is lined by a thin elastic hyaline membrane called the *lamina vitrea* of the choroid. Upon the inner side of this hyaline membrane lies a single layer of large hexagonal cells, containing pigment granules in the whole body of the cell. This is the *pigmentary epithelium* layer, which formally was counted as a part of the choroid and later as a part of the retina. Its cells have brush-like offsets on their

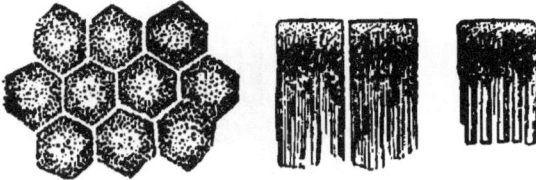

FIG. 10.—(After Schultze). Pigment epithelium cells from the retina of man. To the left a view from their exterior surface. The other cells show the offsets towards the retina which pass in between the rodes and cones.

inner surface which enter between the outer segments of the rods of the retina, withdrawing under the influence of the light and getting prolonged during rest in darkness. The retinal purple, which gives the outer surface of the retina a purplish tint during life, is exuded by this layer, which is therefore to be considered as a special glandular organ interposed between the choroid and retina. (See Fig. 10).

Near the firm attachment of the uveal tract to the sclera, at the corneo-scleral margin, the former becomes rapidly thicker, and thus forms the ciliary body with the ciliary processes on its inner surface. This thickening of the uveal tract is especially caused by the presence of the ciliary muscle. The tendon of this muscle, by its insertion into the corneo-scleral tissue, forms the firm attachment between the uveal tract and the sclerotic. The fibres of this muscle, which are non-striated, spread fan-like backwards and a little inwards in the ciliary body and are finally lost in the choroid.

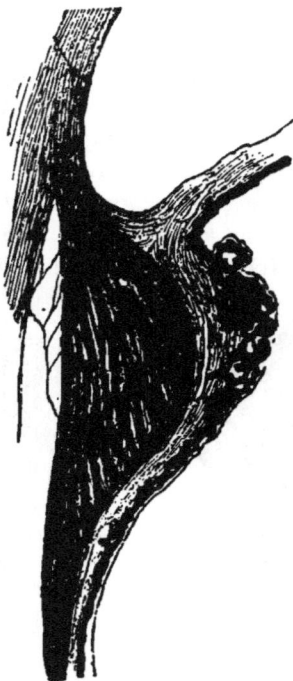

FIG. 11.—Meridional section through the ciliary body of a very short-sighted (elongated) eye. The fibres of the ciliary muscle run almost all in a meridional (longitudinal) direction. The iris-angle is wide.

There are two apparently distinct sets of muscular fibres, the one lying superficially and running in a longitudinal (meridional) direction, the other lying more deeply and running in a circular (æquatorial) direction. The former kind prevails in elongated (short-sighted) eyes (See Fig. 11), while the latter kind predominates in short (far-sighted) eyes. (See Fig. 12).

The ciliary muscle lies embedded in the outer (more superficial) part of the ciliary body. Its inner surface is covered by the tissue proper of the uveal tract with its vitreous lamina. Upon the inner side of the latter lies the thick, dark uveal layer, the continuation of the pigmentary epithelium layer, and further on, on the inner side of this lies one layer of cylindrical cells, which gradually decrease in height toward the insertion of the iris. This layer is considered to be a continuation of the retinal tissue and is called the ciliary part of the retina (*pars ciliaris retinæ*). Since, however, the ciliary body

FIG. 12.—Meridional section through the ciliary body of a very far-sighted (short) eyeball. The inner fibres of the ciliary muscle which run in a circular (aequatorial) direction are very numerous. The iris angle is very narrow.

has of late been more and more clearly demonstrated to be the organ by which the aqueous humor is secreted, and has been directly called the gland of the aqueous humor, it may well be that this so-called retinal layer had better be considered as the glandular layer of the ciliary body.

On the inner surface of the ciliary body, we can distinguish between a posterior, smooth part, the *pars non-plicata*, and an anterior wrinkled part, the *pars plicata*. The folds and wrinkles of the anterior part are caused by small irregular projections called the *ciliary processes.*

These processes, about seventy in number, are directed toward the axis of the eye-ball, and form a circle or wreath upon the inner surface of the globe behind the insertion of the iris.

The arterial blood-vessels of the ciliary body come from the anterior ciliary arteries, which reach the eyeball with the recti muscles, and from the long posterior ciliary arteries; the veins carry the blood back partly into the conjunctival veins and partly into the venæ vorticosæ.

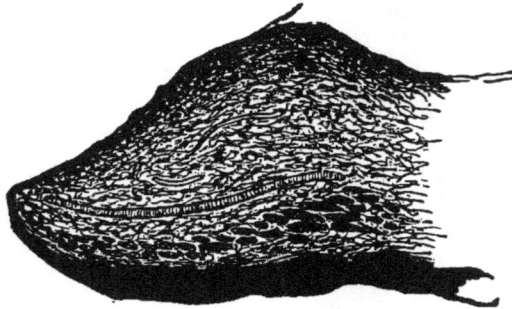

FIG. 13.—Pupillary edge of the iris. Near the posterior (uveal) surface the fibres of the sphincter pupillæ muscle are embedded.

The ciliary nerves form a coarse network on the inner surface of the ciliary muscle and send small branches into it and to the iris.

The *iris* is inserted into the ciliary body just before the tendon of its muscle is merged in the corneo-scleral tissue, and the plane in which it lies forms nearly a right angle with the axis of the eyeball. It forms an adjustable diaphragm across the eyeball and is pierced by a central opening, the pupil.

The bulk of the tissue of the iris is the same as that of the ciliary body. This tissue is in the main a very loose network of fibres and cells, and contains innumerable blood-vessels.

Near its anterior and posterior surfaces, it becomes more
dense, however, and consists largely of spindle-shaped cells.
On its posterior surface it has the darkly pigmented, thick,
uveal layer and on its anterior surface a delicate layer of endothelial cells. The anterior surface of the iris is uneven on
account of a large number of shallow depressions and openings
leading into crypts, dipping for some distance down into the
tissue, and of wrinkles produced by the contraction of the iris-
tissue.

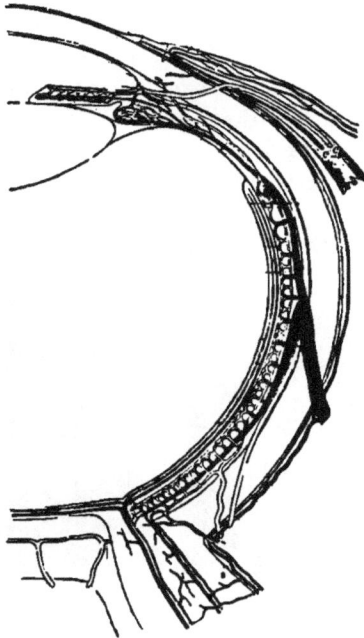

FIG. 14.—(After Leber). Shows the manner in which the arterial blood-supply of
the eye is distributed within it and in which the venous blood is carried
out of the eye. The dark vessels are the veins.

Near the pupillary margin and nearer the posterior surface
of the iris, we find embedded in the iris tissue a ring of organic muscular fibres, the *sphincter pupillæ*, which, by its contraction, reduces the size of the pupil. (See Fig. 13.) Some
authors maintain, also, the existence of an antagonistic mus-

cle, which is said to lie along the posterior surface of the iris and to run in a radial direction; it is called the *dilator* muscle of the pupil. I have never seen it in the human eye.

The arteries of the iris come from a large circular blood-vessel, which lies near the insertion of the iris into the ciliary body, and is formed by the anastomoses of the anterior ciliary arteries. This is called the *large iris-circle*. From this arterial ring branches run toward the pupil in a radial direction. After having formed another ring, the *small iris-circle*, and just before reaching the sphincter muscle of the pupil, they split into a network of capillaries which are distributed between the fibres of this muscle. (See Fig. 14).

The veins of the iris run back to the ciliary body, and finally empty their blood into the venæ vorticosæ.

The arteries of the iris have a much thicker muscular coat than any other arterial blood-vessels of their calibre in the human body.

The nerves of the iris come from the ciliary nerves.

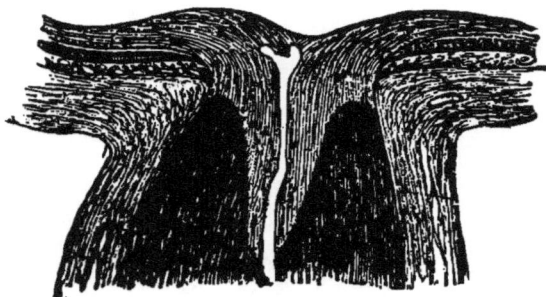

FIG. 15.—Longitudinal section through the entrance of the optic nerve into the eye. The lamina cribrosa is not drawn in order to better show the manner in which the medullated optic nerve fibres are changed into non-medullated ones. This is from a hypermetropic eye; in a myopic eye, the line of transition is convex forwards.

§7. The *optic nerve* when it reaches the sclerotic is enclosed in three sheaths. The outer sheath is formed by the *dura mater* and closely applied to and lining this is the *arachnoid* sheath. These two sheaths become merged in the sclerotic at its posterior surface, and do not enter the eyeball. The third or inner sheath is a continuation of the *pia mater* of the brain. It encloses the nerve directly, and also forms the network of connective tissue in which the fibres of the optic nerve lie embedded. It enters the globe with the nerve, joins the inner layers of the sclerotic and ends in the *lamina cribrosa*, through which the bundles of nerve-fibres pass on into the eyeball.

Just before entering the lamina cribrosa the optic nerve becomes a little thinner, and its nerve-fibres lose their double contour. As soon as the nerve-fibres have reached the inner surface of the choroid, they bend nearly at a right angle with their former direction and expand to form the inner (nerve-fibre) layer of the retina. (See Fig. 15).

After passing through the sclerotic and choroid (lamina cribrosa), and before entering the retina proper, the nerve-fibres form a slight, roundish elevation called the *optic papilla* or *optic disk*. Owing to the manner in which the nerve-fibres thus enter the eye and immediately change their direction, the normal optic papilla shows a more or less centrally located, funnel shaped depression, the so-called *physiological excavation.*

In the center of the optic nerve, and through this funnel-shaped depression, the central retinal artery and vein enter the eyeball, to be distributed exclusively in the retina.

§8. The *retina* is separated from the inner surface of the choroid by the pigmentary epithelial layer. The retina proper reaches forwards to the ciliary body, where it ends with a scolloped edge, called the *ora serrata* of the retina.

The retina is the light-perceiving organ, and has a very complicated structure. If we do not count the pigmentary epithelial layer, it consists of nine distinct layers.

The most external layer is that of the rods and cones, then follows the (doubtful) external limiting membrane. The third layer is the outer granular layer, then comes the outer molecular layer. Then follow the inner granular and inner molecular layers. The seventh and eighth are the ganglionic and the nerve-fibre layers, and these are separated from the vitreous body by the ninth and last layer, the inner limiting membrane.

While the last three layers are called the brain-layers of the retina, the other six are designated as its neuro-epithelial layers. (See Fig. 16).

FIG. 16.—Schematic representation of the layers of the retina. (After Merkel). The inmost layer (downwards) consists of optic nerve fibres which are connected with ganglionic cells (ganglionic layer) which, in turn, send their offsets into the inner molecular layer and there meet with the fibres coming from the cells of the next outer layer, the inner granular layer. These cells send fibres, also, outwards into the outer molecular layer, where their direct connection with fibres coming from the cones has been proven. The cells forming the outer granular layer also send fibres into the outer molecular layer and on their outer side are connected with the rods. The line cutting the bases of the rods and cones is called the outer limiting membrane. The inner limiting membrane, dividing the retina from the vitreous body, is not seen in this drawing.

All the elements of which the the retina consists, are held

together and in position by supporting connective-tissue fibres, called *Mueller's* fibres.

While a direct connection between the optic nerve-fibres and the cones, within the retina, has been anatomically demonstrated, the existence of such a connection between nerve-fibres and rods has so far not been shown. The retinal purple gives its tint only to the outer half of the rods; the cones are untinted.

To the outer side of the optic papilla and slightly below its horizontal diameter lies the yellow spot (*macula lutea*), the point of acute vision. (See Fig. 17). It has a small depression, the

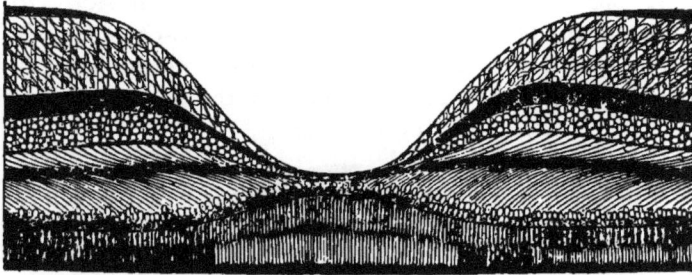

FIG. 17.—(After Helmholtz). Section through the macula lutea and fovea centralis of the retina. Shows that in this part of the retina the rods disappear and the cones stand closer together and are longer and thinner than in other parts of the retina. The nerve fibre layer ends near the fovea centralis. The ganglionic cell-layer is considerably thicker in the periphery of the macula lutea, and almost disappears in the fovea centralis. The same is the case with the outer molecular layer. The remaining layers are so thin within the fovea centralis, that it is but barely possible to recognize them.

fovea centralis, excentrically situated. The retinal tissue at the periphery of the yellow spot is somewhat thickened, while in the fovea centralis it is exceedingly thin. Moreover, in the yellow spot the cones are thinner and longer, and consequently more numerous than in any other part of the retina, and the rods are almost altogether wanting. The proportion between the rods and cones grows in favor of the former from the macula lutea to the ora serrata. These two facts seem to prove that, for distinct vision, the cones are of much greater importance than the rods.

The blood-vessels of the retina, branches of the central
retinal artery and vein, lie chiefly in the nerve fibre and ganglionic layers, but sometimes they reach even into the inner
molecular layer. The outer layers have no blood-vessels
and receive their nutrition from the chorio-capillaris. In the
periphery of the retina the arteries and veins unite to form
terminal loops.

§9. Behind the iris lies the *crystalline lens*, a transparent
lentil-shaped body. It consists of the anterior capsular epithelium, and of the so-called lens-fibres (elongated epithelial
cells), and is inclosed in a hyaline, elastic sack, the lens-capsule.

The part of the *lens-capsule* which lies anteriorly to the
equator of the lens is called the anterior lens-capsule. On its
inner surface lies the single layer of cuboid epithelial cells, the
capsular epithelium. It is thicker than the posterior lens-capsule, which is devoid of epithelium.

FIG. 18.—(After J. Arnold).   Anterior surface of the adult crystalline lens.

The lens-fibres (or lens-bands), which form the main part of
its structure, are also epithelial elements. Where their ends
join each other beneath the lens-capsule, they form sutures,
which are seen to run in a radial direction from the anterior
and posterior poles of the lens, forming angles with each other
of about 120°. On the anterior surface of the lens two of
these sutures run upwards while one runs downwards. On the
posterior surface these conditions are reversed. (See Fig. 18).

The crystalline lens is suspended from the ciliary body by
the *zonule of Zinn*, or *suspensory ligament*. This consists of

tough, transparent fibres, which come from the vitreovs body. While on their way forwards they are bound down to the ciliary body and follow all the depressions and elevations of the ciliary processes until they reach the inner anterior angle of the ciliary body. From here they bend abruptly inwards, and partially crossing each other are inserted on the anterior and posterior lens-capsule, a short distance from the equator.

In the normal eyeball the pupillary margin of the iris rests, and, when moving, slides upon the anterior lens-capsule.

§10. The space between the posterior surface of the cornea, the anterior surface of the iris, and the central portion of the anterior lens-capsule, is called the *anterior chamber*. It contains a clear, watery fluid, without organized elements, the *aqueous humor.*

The space bound by the peripheral part of the anterior lens-capsule, the zonule of *Zinn*, the anterior surface of the ciliary body, its tendon, and the posterior surface of the iris is called the *posterior chamber*. It also contains aqueous humor.

The whole space backwards from the lens and zonule of *Zinn* is filled with a transparent gelatinous substance, the *vitreous body*. This has on its anterior surface a depression, the *fossa patellaris*, in which the lens lies. In the region of the optic disk a small fissure can be traced in the vitreous body from behind forward towards the patellary fossa, called *Stilling's* canal; the hyaline artery lies in this fissure during embryonic life. The vitreous body, especially in its peripheral parts, contains a moderate number of wandering cells.

We have stated above that the sheaths of the optic nerve are direct continuations of the *meninges* of the brain. The inter-vaginal spaces of the optic nerve are in fact in direct communication with the intra-meningeal spaces in the cranium, and can be injected from them.

The course of the fluids within the eyeball is at present thought to be from behind forwards through the vitreous body, the zonule of *Zinn*, and the posterior chamber into the anterir chamber; and the exit of these fluids is thought to take place through *Fontana's* cavities into *Schlemm's* canal and the lymphatics and veins of the sclerotic and conjunctiva.

There are direct communications between the supra-choroidal space and Tenon's space where the venæ vorticosæ and the ciliary arteries and nerves pierce the sclerotic, and fluids may escape by these channels from the eyeball into Tenon's space.

§11. The *external muscles* of the eyeball are six in number. Five of these, the four recti muscles and the superior oblique muscle, together with the levator muscle of the upper eyelid, take their origin from the apex of the orbit around the canalis opticus. The inferior oblique muscle comes from the inner margin of the lachrymal canal.

§12. The *lachrymal apparatus* consists of the lachrymal gland, the puncta lachrymalia, the canaliculi lachrymales, the lachrymal sack, and the nasal duct.

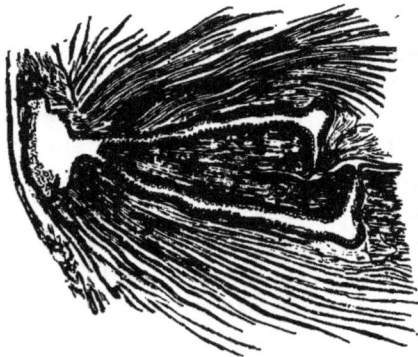

FIG. 19.—(After Gerlach). Shows the direction in which the lachrymal canaliculi run while passing from the lachrymal papillæ through the eyelids to the lachrymal sack.

The *puncta lachrymalia*, as stated above, lie near the inner angle of the palpebral fissure at the apex of the *lachrymal papillæ*, and are the external orifies of the *lachrymal canaliculi*. The latter, after having run into the lid for a little distance at a right angle to the lid margin, turn abruptly towards the nose and converge towards the lachrymal sack. (See Fig. 19). Just before reaching the latter they unite into one short canal. The *lachrymal sack* forms an oblong receptacle for the tears lying behind the *ligamentum palpebrale internum.* Its upper portion (*cupola*) lies higher than the entrance of the canaliculi. The lachrymal sack is about twelve millimeters long, and ends below in the nasal *lachrymal duct.* The latter opens upon the mucous membrane of the inferior nasal meatus, just under the insertion of the inferior turbinated bone.

The lachrymal sack is surrounded by bone at its posterior surface only, the nasal duct is enclosed in bone, except at its nasal extremity. Here the duct pierces the mucous membrane obliquely. Its orifice forms a long oval slit. The length of the duct is about 15 to 20 millimeters.

The *lachrymal gland*, which secretes the tears, is divided into two portions, an upper, larger, and a lower, very much smaller one. The upper portion enclosed in its tough capsule, lies in the lachrymal fossa of the frontal bone, just behind the upper outer margin of the orbit. The lower portion, which consists only of a few loosely connected acini, rests upon the fornix of the conjunctiva, just below the upper one. Small ducts lead the tear-fluid into the conjunctival sack, whence it flows into the nose through the lachrymal puncta, and the remainder of the drainage part of the lachrymal apparatus. Around the base of the lachrymal papillæ lies a minute muscle—*Horner's* muscle—whose function it is to assist in sucking up the tear-fluid.

3. The *blood supply* of the optical apparatus comes from the branches of both carotids. The external carotid reaches the eye from the surface. The ophthalmic artery, coming from the internal carotid, enters the orbit with the optic nerve, lying beneath it within the optic canal. Here it gives off small branches running to the nerve and its sheaths. Within the orbit the ophthalmic artery lies first on the temporal side of the nerve, then it rises and passes between the optic nerve and the superior rectus muscle towards the inner wall of the orbit, where it forms the posterior and anterior ethmoidal arteries. During this passage it gives off the following larger branches, the lachrymal artery, the supraorbital artery, the posterior ethmoidal, and the naso-frontal artery.

The lachrymal artery lies on the temporal side, passes
forward between the rectus superior and rectus externus to the
lachrymal gland, and thence to the eyelid and forms the lateral superior and inferior palpebral arteries which anostomose
with the naso-frontal. The supraorbital artery passes forward
with the nerve of the same name under the roof the orbit and
by way of the supraorbital incision reaches the forehead.
The naso-frontal artery passes forward on the nasal side of
the orbit and after having left the orbit spreads in the neighboring tissues. During their passage through the orbit all
of these blood-vessels give off small branches for the muscles
of the orbit, the orbital fat and connective tissue. (See Fig. 20).

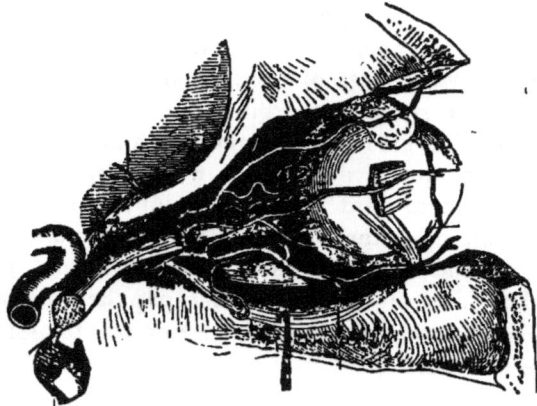

FIG. 20.—(After Merkel). Left orbit uncovered from above to show the distribution
of the arterial blood-supply in the orbit, eyeball and neighboring parts.

The anterior ciliary arteries come from such muscular
branches in the four recti muscles. The posterior ciliary
arteries come either directly from the ophthalmic artery, or
from its larger branches, soon after they have left the ophthalmic artery. There are usually six larger ciliary arteries
which divide each into three or more branches and pass
forward to the eyeball which they enter around the entrance
of the optic nerve. According to their manner of spreading in
the interior of the eye, they are then called short or long posterior ciliary arteries. The central retinal artery comes either

as a separate branch directly from the ophthalmic artery or from one of its larger branches, enters the optic nerve near the eyeball in an oblique direction and with it reaches the retina. The infraorbital artery passes with the nerve of the same name through the infraorbital canal to the surface and running forward through this passage reaches the lower eyelid.

All orbital arteries are very tortuous and thus the movements of the eyeball do not interfere with the blood flow.

The finer veins of the orbit follow in a general way the course of the arteries. The superior ophthalmic vein collects the blood from the lids, forehead and lachrymal apparatus, furthermore from the ethmoidal veins, a number of muscular and ciliary veins and the central retinal vein. It passes out of the orbit by the superior orbital fissure, and empties into the cavernous sinus. The lower ophthalmic vein collects the blood from a number of muscular veins, the ciliary veins, and, also going through the upper orbital fissure either joins the superior ophthalmic vein or enters the cavernous sinus separately.

The veins of the eyelid form near the inner canthus the angular vein which joins the anterior facial vein, and near the outer canthus the temporal and facial veins.

§14. Aside from the optic nerve, the orbital cavity contains a large number of smaller *nerves*. They enter the orbit through the superior orbital fissure. The trochlear nerve passes forward with and spreads within the superior oblique muscle. The abducens nerve goes to the external rectus muscle. The oculomotor nerve, soon after having entered the orbit, is divided into two branches. The upper and thinner one forms a branch for the superior rectus, and one for the levator muscle of the upper lid. The lower and thicker branch is again split into three branches for the internal and inferior rectus and the inferior oblique muscles, respectively. The last named branch forms the short root of the ciliary ganglion. The ophthalmic nerve, which takes its origin from the fifth (trigeminus) nerve forms three branches within the orbit, the supraorbital, lachrymal, and naso-ciliary nerves. The supraorbital nerve runs forward under the roof of the orbit, lying on top of the orbital fat (with the artery of the same name), and splits in two branches of which one retains the name, the other is called the frontal branch. The lachrymal nerve goes to the lachrymal gland. The naso-ciliary nerve forms the long root for the ciliary ganglion and then passing to the nasal side of the orbit gives off two or three long ciliary nerves and ends in two branches, the infratrochlear nerve and the ethmoidal nerve, which leaves the orbit through the foramen ethmoidale anterius.

The ciliary ganglion consists of motor, sensory and sympathetic fibres. The motor fibres reach it by the short root formed by the branch of the oculomotor nerve for the inferior oblique muscle. The sensory fibres come from the naso-ciliary branch of the ophthalmic nerve (trigeminus) and from its long root. The sympathetic fibres come from the plexus of the cerebral carotid and join the long root of the ciliary ganglion. From this ganglion spring the short ciliary nerves (from 2 to 6) which divide into about twenty smaller branches, and enter the eyeball in a circle around the optic nerve entrance together with the long ciliary nerves from the naso-ciliary nerve.

CHAPTER II.—METHODS OF EXAMINING THE EYE.

§15. For all examinations of the eye good light is absolutely required. In day-time it is therefore best to see the patient near a window and opposite to it, avoiding, however, bright sunlight. After the patient has been properly seated, a systematical examination should begin with the inspection of the cutaneous surface of the *eyelids*. Then the eyelashes and their position, the orifices of the *Meibomian* glands, the motility of the eyelids, and the size of the palpebral fissure should be carefully noted. To get a good view of the outer and inner canthus, the puncta lachrymalia and the caruncula lachrymalis, it is best to slightly raise the upper eyelid with the forefinger, while the thumb of the same hand gently pulls down the lower lid. This little manipulation, which has to be used very frequently in examining eyes, should be executed without exerting the slightest pressure on the eyeball. If it is impossible to make a perfect inspection with the aid of this manipulation, it will be best to draw the upper eyelid upward with the thumb of one hand and the lower eyelid downward with the thumb of the other hand. If the skin of the lower eyelid is too slippery for this manoeuvre, a towel or piece of linen cloth wound around the thumb will be of great assistance. In thus separating the eyelids all pressure upon the eyeball must be carefully avoided. This is most surely accomplished by laying the thumbs on the skin of the eyelids near the orbital margins and drawing them apart by dragging on the skin only.

If there is any complaint about *stillicidium lachrymarum* (tear-dropping, lachrymation), the first point to be examined into, is, whether the puncta lachrymalia lie in contact with the ocular conjunctiva near the caruncle. Then making pressure on the lachrymal sack, while the puncta lachrymalia are closely watched, the escape of fluid into the conjunctival sack will give us an indication of any obstruction to the proper drainage of the tears into the nose.

If there is an escape of fluid from either punctum, its character, whether watery, mucous or purulent, will be of importance with regard to the diagnosis of an inflammatory process in the lachrymal sack. The further exploration of the lachrymal sack and duct by means of probes will be detailed in Chapter IV.

To inspect the *ocular conjunctiva* we draw the lids apart in the manner just described, and notice whether there is any abnormal condition. If there is hyperæmia, we should make sure whether this hyperæmia is confined to the conjunctival blood-vessels, or whether it involves also the ciliary blood-vessels in the sclerotic near the cornea-scleral margin. This is best done by sliding the conjunctiva slightly upon the sclerotic by means of the eye-lids. A hyperæmia confined to the moveable tissues concerns the conjunctival blood-vessels only. These vessels are, moreover, comparatively large and convoluted, and are easily distinguishable as separate vessels, whereas the deeper-lying ciliary vessels are much finer and appear rather as a ring of diffuse redness with a bluish tint, densest next to the cornea and shading off into the sclerotic.

If the symptoms complained of refer to the conjunctiva of the eyelids, or if a foreign body has entered the conjunctival sack, we must next inspect the *inner surface of the eyelids*. The conjunctival surface of the lower eyelid and the lower fornix of the conjunctiva are easily exposed to view by directing the patient to look upward and drawing the skin of the lower eyelid downward toward the cheek with the thumb. In deeply set eyes, the lower fornix is most perfectly exposed by drawing the lower lid downwards, while the patient also looks downwards (*Arlt.*) The exploration of the conjunctival surface of the upper eyelid and the upper fornix of the conjunctiva requires more skill, and is accomplished in the following way: Place the thumb of the right hand (when examining the patient's left eye) against the orbital margin above the outer angle of the palpebral fissure, then take hold of the cilia with the thumb and forefinger of the left hand and direct the patient to look downwards. Next draw the eyelid, thus held by the cilia, gently downwards and forwards, at the same time shifting the thumb of the right hand into the depression, which appears between the eyebrow and the tarsus, and lastly turn the lid-margin upwards, using the right thumb to keep the upper edge of the tarsus in position, while the whole tarsus is being turned around its upper edge as a fixed center. Instead of the thumb a smaller round object (such as a probe, a pencil, or a match), may be used to fix the upper edge of the tarsus. Examining the patient's right eye, the hands should be reversed. When the eyelids are very forcibly shut and the patient is unable to assist in looking down, or when the conjunctival sack is considerably shrunken, it may be very difficult to bring the conjunctival surface of the upper lids to view. When the eye-lashes are absent, it is often sufficient to direct the patient to look strongly downward, to lay the end of a probe along the upper edge of the tarsus so as to press it gently downwards and backwards, and to draw the lid margin upwards by means of the ball of the thumb applied to the dry skin of the eyelid near its free margin (*Desmarres*). In this, as in the following manipulations, we will be materially aided by first instilling one or more drops of a 4% solution of muriate of cocaine into the conjunctival sack.

In young children the inspection of the lids, as well as of the eyeball, is best affected by taking the child's head in the lap, or, if necessary, between the knees, while its legs rest on the lap of another person. Sometimes a general anæsthetic may be necessary.

When the lids and conjunctiva have thus been explored, we next inspect the *cornea*. A healthy cornea is perfectly transparent and polished, and reflects the light like a mirror. These two peculiarities allow us to distinguish all affections of this membrane easily.

If there is any form of inflammation, or an abrasion, a scar, or a foreign body present, the tissue of the cornea will be seen more or less affected, either in its transparency, or in the perfection of polish of its surface. Any considerable changes in the curvature of the surface of the cornea will be easily detected by putting the patient in such a position that we can see the reflected image of a window or a flame on the cornea.

By then directing the patient to move his eye, say in a horizontal direction, so as to allow the reflected image to move, so to speek, over the cornea, it will become distorted as soon as it reaches the part in which the curvature is altered. When the injury is so small that we can barely find it, staining with a solution of fluorescein, 10 grains to the ounce, is of great value. While the normal portion of the tissue remains uncolored, the parts denuded from epithelium take on a green tint.

The sensibility of the cornea is examined by touching it with a camel's hair brush or a small roll of tissue paper.

After examining the cornea, we inspect the contents of the anterior chamber, the *aqueous humor*, which, in the normal condition, is also perfectly transparent. Any lack of transparency in this fluid, is due to an affection of the deeper portions of the eyeball.

In examining the *iris*, we have first to pay attention to any anatomical changes in its tissue, and then to its function as a moveable diaphragm. The pupil ought to expand promptly on shading the eye, and to contract promptly on exposing the eye again to the light.

If the iris is inflamed, there is hyperæmia of both the conjunctival and ciliary blood-vessels. The latter show as a pink or bluish-red zone around the corneo-scleral margin and are not moveable with the conjunctiva. The tissue of the iris appears swollen and loses its lustre. The color of the iris is also changed, in blue eyes taking on a greenish shade, in dark eyes a dirty brown. After iritis has become established, the pupil is nearly or wholly immoveable.

The motility of the pupil may also be disturbed when there are no inflammatory symptoms present. In order to test this, we cover the healthy eye so as to exclude all light from it, and then, alternately shading the other eye with the hand and exposing it to light again, we watch the size of the pupil. While this examination goes on, the patient must keep his eyelids well apart and look steadily in the same direction. If the pupil remains unchanged, under the influence of alternate light and shade, we must see whether it contracts, perhaps, during the effort to accommodate for a near object.

We should, furthermore, see, whether the iris trembles when the eyeball is moved. The size and shape of the pupil, when at rest, are also to be noted.

The position of the plane of the iris is also of importance. We must see, whether its periphery is bulged forwards or drawn backwards, or whether any particular part of it is protruding, etc.

If the pupil is immoveable, or acts imperfectly, it is best to test its dilatability by the instillation of a mydriatic. The simplest one for a mere examination is a one per cent. solution of homatropinum hydrobromatum, as its action disappears very readily (in from 8 to 12 hours). If a stronger mydriatic is needed, as is generally the case in inflammation of the iris, a one per cent. solution of atropinum sulfuricum should be used in its place.

An inflammatory process of the *ciliary body* is recognized by a deep bluish-red zone of injection around the corneoscleral margin, and by pain on pressure upon the ciliary region. The latter symptom is easily ascertained, by pressing slightly upon the ciliary region through the closed eyelids with a pencil or any rounded small object, or even with the finger. (Tenderness of the ciliary region on pressure may also be present in iritis).

Inspecting the *crystalline lens*, we have chiefly to notice the transparency of its capsule, cortex and nucleus. In order to see the equatorial portions of the lens the pupil must be widely dilated. If the lens is wanting, or is dislocated, from the patellary fossa, the iris will tremble (*iridodonesis*), except when the whole lens lies in the anterior chamber, a condition which presents otherwise characteristic appearances. See Chapter XV.

If the lens is transparent enough, we may also be able to see a cyclitic membrane, changes in the anterior portion of the vitreous body, a detached retina, or an intra-ocular tumor through it. For the examination of the anterior third of the eyeball, we make use of the so-called *focal or oblique illumination*, which enables us to detect, for instance, slight changes in the cornea or lens, which in the diffuse illumination may escape our notice altogether. The patient is seated opposite the examiner and a lamp is placed at the side and somewhat in front of the eye under examination, a convex lens of two and a half or three inches focus, is used to throw a pencil of light obliquely upon the parts under examination. By moving this lens nearer to or farther from the eye the focus may be thrown upon deeper or more superficial parts. It is sometimes a decided help in making a diagnosis to view the parts thus illumi- • nated with a magnifying lens held in the other hand. (See Fig. 21).

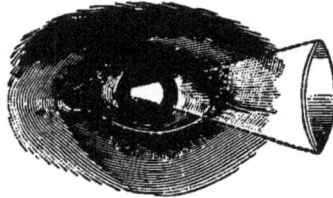

Fig. 21.—(After Meyer). Focal or oblique illumination of the anterior parts of the eyeball.

Tumors, abscesses, etc., within the *orbit* may be detected by the protrusion and displacement of the eyeball and can often be located by palpation.

§16. Thus far, we have treated of the visible changes occurring in affections of the anterior part of the eyeball only. Eye affections, not accompanied by changes, visible to the naked eye, call for the testing of vision subjectively.

The *acuteness of vision* is tested by means of test-types, and when it is very much reduced, by the outstretched fingers of one hand. The test-types in use are constructed in such a manner that their limbs are seen by the normal eye under a visual angle of one minute, while the whole letter is seen under an an angle of five minutes. Of these letters one set is used for distant vision and contains letters to be seen by the normal eye distinctly at from 200 to 20 or less feet distance. The other set is constructed for near vision. The acuteness of vision is expressed in the form of a fraction, the denominator of which gives the distance at which the letters ought to be recognized, while the numerator gives the distance at which they are actually seen. The normal eye must see the letters called XX, in our set of test-types, at 20 feet, and this is noted in the following way, V (visus)$=^{20}/_{xx}$.   If the patient sees the letters, which the normal eye recognizes at 100 feet, at 20 feet only, we write V $=^{20}/_c$, and his visual acuteness is said to be only one-fifth of that of a normal eye.   In making such examinations the test-types must be well lighted and the patient must sit or stand with his back to the light.

When no letters can be distinguished, we may examine the acuteness of vision by means of the outstretched fingers.   In doing so, we should be careful to hold the fingers against a dark back-ground and move from the distance towards the patient until he can count them.

§17.   The whole region within which an eye, when perfectly at rest, can perceive objects, is called its *visual field*.   The examination of the visual field is best made by an instrument, called perimeter, of which a number of patterns are in use, consisting of an half or quarter arc which can be moved in the directions of all the meridians.   A white or colored object is slid along this arc from the periphery towards the center, and the patient who fixes the center announces when he sees this object.   The field can then be drawn on forms printed for the purpose.   (See Fig. 22).   It can also be directly projected on a plane surface (black-board) and thus be accurately mapped out.   The simplest method is, to let the patient cover one eye and to direct him to gaze steadily, with the other one into the observer's opposite eye.   Then move the fingers, or a small staff with a white tip, from different directions towards the line connecting his eye with that of the observer, keeping always at an equal distance from both, and notice when he first recognizes it.   If your own eye is normal and gazes steadily into the patient's eye during this procedure, the extent of your own visual field will allow you to notice at once any defect in his.

If the patient's sight is so poor, that he can no longer recognize the fingers, or other small objects, as in a case of cataract, the visual field is best examined with a candle-flame in a dark room.   This is done exactly in the same way, directing the patient to look straight ahead and not to change the position of his eye and moving the candle towards the visual line in the direction of the different meridians.   Care should be taken to shade the patient's eye whenever the direction of the candle is changed.

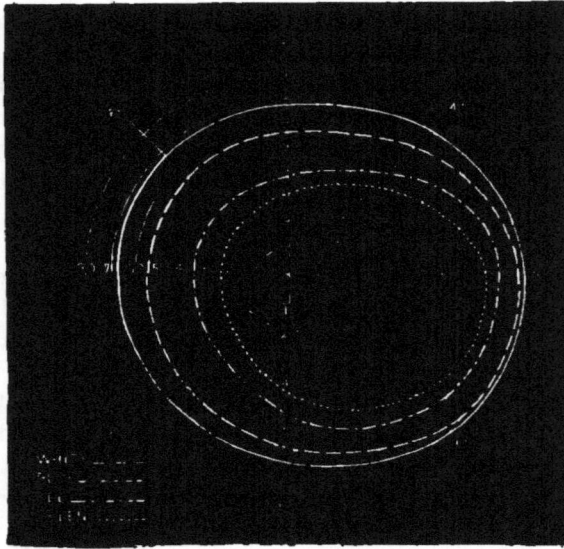

FIG. 22.—(After Landolt). Outlines of the visual field of the right eye for white blue, red and green, showing at the same time the manner in which the outlines of a pathologically changed visual field can be mapped out.

§18.   The *light-sense* is best examined by Foerster's photo-meter, an instrument in which the light of a candle (of one candle-power) illuminates stripes of black and white on the the opposite wall of a blackened box, into which the patient looks. The size of the window through which the light is admitted can be changed at will and thus the lowest amount of light which allows of a differentiation between the white and black stripes in a given case can be easily found. By then comparing the result with that of an eye of a normal power of differentiation we are enabled to tell whether the patient's light-sense is normal or diminished.

§19.   Sometimes we have to examine a patient in regard to

his *perception of colors.* A great many methods for this purpose have of late come into use, since the subject of color-
blindness has received special attention in connection with the
marine and railroad service of almost all civilized countries.
Holmgren's method, in which .skeins of variously colored
wool are employed, is the most convenient. The patient is
first shown a light green skein and asked to match it with similar tints. If he is color-blind, he will make strange mistakes,
selecting gray-green, brown, yellowish, pink and grayish red,
etc. For further details, see chapter XXIV.

§20. The *intra-ocular tension* is best examined by directing
the patient to look down, and and then gently laying the tips
of both index fingers upon the upper lid, and alternately pressing them upon the globe, as we are accustomed to do in
searching for fluctuation. We determine in a general way,
whether an eye is harder or softer than normal, by comparing
it with its fellow. When both eyes are affected, the tension
should be compared with that of the healthly eyes of another
person, unless the examiner is expert enough to judge acur-
ately without such comparison.

§21. The *accommodative power* of an eye is examined by
directing the patient to look at a small object (finest test-
types) and moving it so close to the eye that he can but just
recognize it. If the accommodation is defective this near-
point (*punctum proximum*), will be farther from the eye than it
should be, taking into account the age of the patient. (See
Chapter XX).

This examination will at the same time give us a hint with regard to the state of *refraction* of the examined eye. If it can
read finest print for a prolonged period and at a smaller distance from the eye than the age of the patient would warrant,
the eye is short-sighted. If it is unable to read the smallest
print at all or for any length of time, at the normal distance it
is probably far-sighted or astigmatic.

§22. In order to see any changes in the conditions of the
posterior portions of the eye, of the vitreous body, the optic
nerve, the retina and the choroid, we have to make use of the
ophthalmoscope.

Light thrown into an eye will not only be preceived there, but it is also reflected. The reflected rays return to the source of light by the same way by which they have entered the eye.

When the pupil of an examined eye is very large the observer's unaided eye is sometimes able to catch such rays returning to their source, and then he sees the usually black pupil appearing shining red. In this way, of course, no details of the background of the eye are to be distinguished. To make this possible it is necessary to bring the observer's eye into the axis of the returning pencil of light-rays. This is done by throwing light into the eye by means of a mirror, perforated by an opening, through which the observer looks. Armed with such a mirror, with suitable correcting glasses behind the central opening, the observer's eye is enabled to view all the details of the background of the examined eye and even to measure the refraction.

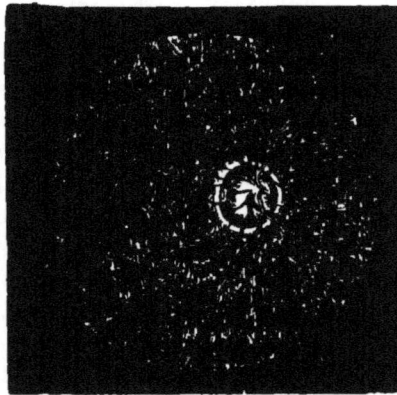

FIG. 23.—(After Jaeger). Normal appearance of the fundus of an eye. Arteries light, veins black. The blood-vessels of the choroid shining through the retina.

In examining eyes with the ophthalmoscope, we make use of two different methods, called the direct, and the indirect method.

In the direct method the observer's eye, armed with the mirror, is brought as near to the examined eye as is possible, without excluding the light. The image seen by this method is the virtual erect image of the background of the examined eye.

In the indirect method, the eye armed with the mirror, is moved from the examined eye to a distance of $1^1/_2$ or 2 feet, and a lens of from 2 to 3 inches focus is held before and within about 2 inches of the latter. The observer's eye now catches the real inverted aërial image of the background of the examined eye at or near the focus of the objective lens.

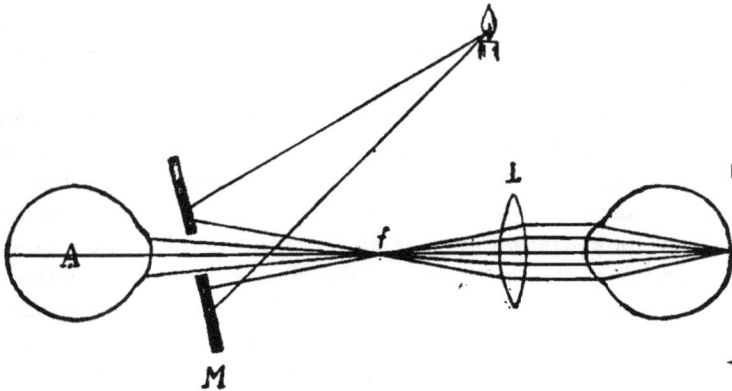

Fig. 24.—Explains, how rays of light from a candle reflected by a mirror (M) with a central perforation are thrown into the eye to be examined. The rays of light returning from the eye under examination are brought to a focus (f) by a bi-convex lens, of about three inches focus, where they form an inverted image of the fundus of the examined eye which is seen by the observer (A). This is called the indirect method (inverted image).

In the indirect image the details seen are smaller, but the field is larger, in the direct image the field is small, but the details are much larger.

In both methods we may use either artificial light or diffuse daylight. The former is more convenient, and is generally employed.

By means of the ophthalmoscope we are enabled to scan very closely the largest part of the back-ground of the eye and detect anything abnormal. The ophthalmoscope may further be used to examine the transparency of the anterior parts of the eye and for the determination of errors of refraction, as stated.

The interior of the eye may also be inspected directly when eliminating the refractive power of the cornea, by placing it under water, by pressing a plain glass plate against the cornea and thus flattening it (*Bellarminoff*), or by means of a plano-concave meniscus placed in front of the cornea, with the concave side touching it (*Koller*) after having been moistened.

§23. We should further examine into the *motility* of the eyes, especially noticing if the movements of one or both eyes are restricted or excessive in any direction. This mode of examination is mostly called for in cases of strabismus and in paralysis of one or more of the external muscles.

Direct the patient to fix his gaze upon your forefinger, and while moving it toward and from his nose in the middle line, observe whether his binocular fixation is preserved within the whole range of his accomodation. Next cover one eye and let him look straight at your finger with the other, then quickly removing the cover from the first eye, note whether it makes any movement to come back to the point of fixation. Then direct him to follow your finger in different directins with both eyes and pay particular attention to the excursion of each eye.

If one or more muscles of one eye (or both) refuse to act or act to an undue degree *double vision* (*diplopia*) must result, as binocular vision is no longer possible. The patient will see a true image with the healthy eye and a false one with the diseased eye.

The examination for double images is best made with a candle flame. This is moved before the patient's eye at a distance of six or eight feet in all directions, and the patient is directed to say when he sees double. In order to enable him the better to distinguish the second image, one eye, usually the healthy one, is armed with a colored glass.

Particular attention has sometimes to be paid to the action of the *internal recti* muscles during convergence, as in some people these muscles refuse continued work. Moving the forefinger towards the patient's nose, while his gaze is fixed on it, will often enable us to detect such a weakness, one eye presently diverging. To make sure of such an observation, it is then best to let the patient look at a small object [a line with a dark dot in the center (*v. Graefe*)] at reading distance, while a prism with base up or downwards is held before the suspected eye. If the false image thus produced does not stand directly above or below the real one, but stands also to one side, an insufficient action of one or both recti interni is shown. The prism, by means of which, with its base inward, we can bring the two images into the same vertical line, gives us the degree of insufficiency. However, other muscles besides the interni may be insufficient.

For the different forms of insufficiency of the different external muscles of the eye the terms of *esophoria* (tendency to overconverge), *exophoria* (tendency to overdiverge), and *hyperphoria* (tendency to move too far upward) have been brought into use by *Stevens*. To these are added the terms of *hyperexphoria* and *hyperesophoria* when one or the other of the oblique muscles is also affected. These more minute insufficiencies (in general called heterophoria) which by some are thought to be of grave importance may be best inquired into by means of a *Maddox* glass-rod, *Stevens'* phorometer or *Savage's* double prism, or some similar contrivance by which the abnormal position and direction of the image of one eye is made more perceptible, by comparing it with the position of the image of the fellow eye.

CHAPTER III.—DISEASES OF THE EYELIDS.

§24. The skin of the eyelids may be attacked by all forms of *skin disease*. Among these, however, several are more commonly found on the eyelids than others.

Erysipelas attacks the eyelids usually during an attack of erysipelas of the face. It must then be treated conjointly with the latter. Local applications of ichthyol over the infiltrated parts and somewhat over the neighboring healthy skin with a brush are highly to be recommended.

Herpes Zoster Ophthalmicus is that form of herpes in which the vesicles are found in the skin in the neighborhood of the eye (and sometimes on the eyeball itself), where the terminal fibres of the trigeminus nerve are situated. Their appearance is usually preceded or accompanied by violent pain. The progress of the disease is the same as in other regions. The contents of the vesicles, at first watery, become soon purulent, and finally a crust is formed under which the small ulcer heals, leaving a lasting scar. A drying powder may be applied with benefit. No other treatment is apparently necessary or of any value.

Chronic Eczema, when found on the lids, is best treated by an ointment containing oxide of zinc.

Varioloid or true *variola* pustules, are found on the lids during a general attack of these diseases. Peculiar pustules are produced by an infection with animal-lymph, and have been termed *vaccinola*.

Œdema of the lids may be produced by a variety of conditions which interfere with the current of the venous blood, such as conjunctival diseases, inflammation of Tenon's capsule, orbital phlegmon, inflammatory processes in the interior of the eyeball, orbital tumors, or inflammatory processes in and around the lachrymal sac. In these cases the œdema of the lids is purely symptomatic.

—14—

In rare cases an œdema of one or both lids of one eye, or of both eyes is seen to take place without any known cause. Such an œdema generally disappears as it has come, without any interference.

In some patients the exhibition of large doses of some iodide salt causes an acute œdema of the lids, which disappears as soon as this remedy is withdrawn.

§25. The most frequent affection of the lids is confined to the lid-margins and is called *blepharitis ciliaris or marginalis* or *blepharadenitis.*

The prominent symptoms of this affection are the formation of scales or larger crusts along the lid-margin, at the roots of the eye-lashes, also redness and swelling, which latter are usually confined to the lid-margin.

When the disease progresses, the original small scales are replaced by larger crusts, in which the eye-lashes are often totally embedded. The swelling increases sometimes to such an extent that the lid-margin is turned outward from the eye, thus giving rise to an ectropium, especially of the lower eyelids. Such an inflammation cannot exist for a long period without affecting the eyelashes also. They fall out, their bulbs become atrophied, and when the inflammation has finally passed away, the eyelids remain more or less destitute of eyelashes (*madarosis*). Blepharitis ciliaris is mostly, in its severer or chronic forms always, combined with some degree of catarrhal conjunctivitis.

From the beginning the disease causes a disagreeable feeling of heat, irritation and weakness, when the eyes are used for small objects. In the morning the eye-lashes are glued together by the dried secretion.

The affection is chiefly one of childhood, although it is observed in adults also. Children of a strumous habit and of a fair complexion are perhaps oftenest subject to it.

The origin of the disease is probably an infection by microorganisms, due to rubbing the lid-margins with dirty fingers or a handkerchief soiled with nasal secretions.

We may distinquish between a *blepharitis squamosa* and a *blepharitis ulcerosa*. While in the former the skin beneath the scales is red and swollen, but otherwise more or less intact, in the latter form we find small ulcers and abscesses under these scales. These ulcers and abscesses are due to pyogenous infection of the hair-follicles and sebacious glands of the lid-margin.

Blepharitis ciliaris does not, as a rule, yield easily to treatment, and in its worst forms a restitutio ad integrum is almost impossible. Where a strumous habit exists, internal treatment should always be combined with the local one.

The treatment which yields the best results in mild cases of blepharitis, is the following: Bathe the eyelids with luke-warm water until the scabs are well soaked. They can then readily be removed by rubbing along the lid-margin back and forth with a dry, rough towel. If luke-warm water does not seem to soak them sufficiently, the application of white vaseline will do so. When all the scabs have been carefully removed, apply a small quantity of an ointment, containing from 2 to 4 grains of yellow oxide of mercury to 3 or 4 drachms of white vaseline or lanoline, or 10 grains of aristol to 1 drachm of white vaseline or lanoline. This is to be rubbed into the roots of the eyelashes while the eyelids are kept closed. After having allowed it to remain there for a few minutes, the surplus ought to be gently wiped off. This application must not be made just before the patient retires, but several hours earlier. If any of the ointment gets accidentally into the conjunctival sack it may cause considerable smarting, but it will do no harm. If the catarrhal conjunctivitis is at all pronounced it should also be treated. (See Chapter VII).

In severer, especially the ulcerous, cases where large crusts glue the eye-lashes together and cover an ulcerated lid-margin, it is best to soften the crusts with white vaseline. We are often obliged to remove such crusts with the forceps. This should be done very carefully and gently. When we have succeeded in thoroughly removing the crusts, the ointment of oxide of mercury should be applied. It will, however, but rarely suffice in these forms of blepharitis, and we are often compelled to resort to a caustic treatment. The application of a 2 or 3 per cent. solution of nitrate of silver with a camel's hair brush, while carefully shielding the eyeball, is highly to be recommended. In some cases the solid nitrate of silver stick must be used. In other cases tar or oleum rhusci, either pure or mixed with vaseline, is very useful. Epilation of the diseased cilia, and burning the ulcerations by means of actual or galvano-cautery may often be used with great success.

Such applications should not be discontinued at once, when the ulceration is healed, but be continued for a longer period, until all swelling and irritability of the tissues of the lid-margin have disappeared.

Such treatment, to ensure perfect success is, at least in the severe cases, best applied by the surgeon himself. He may, however, be in a great measure assisted by the patient. The patient ought to bathe his closed eyelids frequently with cold water, or apply cold compresses to them. He must refrain from using his eyes for all occupations, which are likely to irritate them, especially by artificial light. Children should be kept from school, and be given a separate clean handkerchief to wipe their eyes with. Adults should not smoke in a close room or stay in rooms where others are smoking. Fresh air is not injurious. If blepharitis ciliaris is observed in a patient whose eyes are ametropic (show an error of refraction), and especially, it they are hypermetropic or astigmatic, the correction of the ametropia by glasses will have a beneficial influence on the result of the treatment.

§26. *Phthiriasis*, an affection of the lid-margin which may simulate blepharitis ciliaris, is caused by the presence of crab-lice (*pediculi pubis*) upon the eyelashes. The patient, who is usually unaware of their presence, feels a great irritation on the lid-margin, and by repeated scratching, often produces such a condition of the lid-margins as may easily be mistaken for blepharitis ciliaris. On closer inspection the eggs of the parasite are seen adhering to the eyelashes, and the parasites, themselves, may be recognized burrowing into the openings of the hair follicles.

The treatment consists simply in the application of some mercurial ointment. Among them the common blue ointment is as good as any. About a quarter of an hour after this has been rubbed into the affected lid-margins, the parasites will have come out of the glandular orifices, into which they have burrowed their heads, and can then be easily removed with the forceps. The application may be repeated if necessary. The eggs ought to be seized singly between the teeth of the forceps and gently pulled along the eyelashes, to which they are adherent. This is the easiest method of removing them, and much better than cutting off the eyelashes. As soon as the irritating cause is removed, the inflammatory symptoms disappear.

§27. The acute inflammation of the orifice of a *Meibomian* or tarsal gland, and later of the gland itself, is called *hordeolum*, commonly stye. It begins with a slight circumscribed redness and swelling at the lid-margin (*hordeolum externum*), or farther back on the inner side of the eyelid (*hordeolum internum*) which is often exceedingly painful. The swelling gradually increases and may lead to œdematous swelling of the whole affected lid, so that it may appear like a very serious inflammation. Soon, however, the swelling comes to a head at the orifice of the gland, or on the conjunctival surface of the lid, and, if it breaks and the pus is discharged, the inflammatory symptoms will subside. The result may be hastened by hot fomentations.

The best way to treat a stye is to split the swelling in its beginning with a narrow knife or lancet in a direction at right angles to the lid-margin. The depletion and consequent decrease of tension in the affected parts, and sometimes the removal of an actual obstruction, will cut short the inflammatory process and the patient's suffering.

A weakened constitution or strumous habit seems often to be a predisposing cause of this affection, and such patients sometimes suffer habitually from styes. Tonic treatment is therefore sometimes indicated.

The main cause, however, is probably a direct infection of the Meibomian glands by micro-organisms due to rubbing with a dirty handkerchief or finger. This fact explains, also, in a simple manner, how it happens, that when once one stye has appeared, more are likely to supervene.

When the pus and detritus filling the infected Meibomian gland are not removed by bursting on the conjunctival surface or by the surgeon's knife, a tumor results which is called *chalazion* or tarsal tumor. The acute inflammatory symptoms may have subsided before it was or when it is noticed, and may sometimes be rekindled during its formation. As a rule, however, the formation of such a cystic tumor is comparatively painless, and so long as it remains small it may cause no inconvenience.

Such tarsal tumors sometimes disappear without surgical interference by absorption, and then leave no trace behind. In most cases, however, they remain stationary, or even grow steadily. When they have attained a considerable size, and especially when they lie in the upper eyelid and near the lidmargin, they are not only disfiguring, but cause disagreeable symptoms, and become very annoying. They may even obstruct a part of the visual field.

The contents of the larger tarsal tumors are usually no longer fluid or only partially so. They become organized, and form granulation tissue, which is enclosed in a dense connective tissue sack. In some cases these tumors are firm and of a fibrous or even enchondromatous character.

The best way to deal with the tarsal tumors is to enucleate them. This is done by means of a horizontal cut through the cutaneous surface of the eyelid. This operation may be made bloodless by the use of *Knapp's* lid-clamp. Such a cut will leave no visible scar. Only a clean and perfect removal of the whole cyst-wall will afford security against a relapse.

Some surgeons prefer to remove tarsal tumors from the conjunctival surface of the eyelid, and by means of a sharp spoon.

The smaller tumors, when of a soft consistency and before a tough cyst-wall has been formed, may be opened from the conjunctival surface of the eyelid by an incision at right angles to the lid-margin and parallel with the ducts of the *Meibomian* glands and then effectually squeezed out by means of *Ayres'* chalazion forceps or between the fingers.

§28. *Phlegmonous abscesses* in the subcutaneous tissue of

the lid are comparatively rare. They cause redness, heat and swelling, and fluctuation soon can be felt. As soon as the diagnosis is secured, a knife should be plunged into the swollen part in a horizontal direction, and the pus thus be evacuated.

Syphilitic ulcers, primary as well as from constitutional syphilis, have been observed on the eyelids. They call for no other treatment than the manifestations of syphilis do in other parts of the skin.

Warts and horny excrescences on the lids are of little importance, and may be simply cut off with scissors.

Xanthelasma is a yellowish or brown tumor of the skin. It usually lies near the inner margin of the orbit in the integument of the upper eyelid. It appears often in symmetrical spots on both upper eyelids and forms only a slight elevation. This growth is perfectly harmless, but if the patients, mostly females, desire its removal, a clip of the scissors will easily accomplish it.

Sarcomatous growths are but very rarely observed in the eyelids as a primary affection, but *epitheliomata* quite frequently originate in this region. They appear generally on the lower eyelid, near one of the angles, more frequently on the inner angle, of the palpebral fissure.

These malignant tumors often take their origin from a pre-existing wart, or, if not, they resemble such a harmless growth very much in their early stages. Gradually the wart becomes somewhat sore on the surface, and a little scab is formed, soon grows, and when removed, reveals an ulcerated surface underneath. The tumor slowly spreads and eats away more and more of the lid-margin, and it gradually produces an irregularly shaped, nodular, hard swelling of the adjacent tissue. The eversion of the lid-margin caused by its presence, or perhaps the destruction of the lachrymal canaliculus, allows the tears to drop continually. This and the irritation from the partially unprotected state of the eyeball, become more and more annoying. Sometimes a very distressing shooting pain accompanies the growth of the tumor.

In case it is not interfered with, the epithelioma may extend to the ocular conjunctiva, and thus an epitheliomatous

symblepharon may be formed. In this way the newformation may even enter the interior of the eyeball and spread
there.

The growth of the tumors is slow, and a patient may suffer
from them a very long time before they attain a fatal development, if this ever happens.

The only treatment which promises a radical cure in sarcomatous or epitheliomatous tumors of the eyelids is their early
destruction by galvano-cautery, or their removal by excision.
This latter operation must of course be done, according to
the general surgical rules for the removal of malignant
tumors.

According to the size and situation of the newformation,
its removal will cause a more or less important loss of substance of the affected eyelid, which my have to be made good
by means of a plastic operation. In most cases a part of the
lid-margin and a piece of healthy eyelid will be left after the
removal of the tumor, and these should be carefully made use
of.

In such cases I consider the method for repairing the loss of
substance by sliding flaps (*Knapp*) (See Figs. 25 and 26),

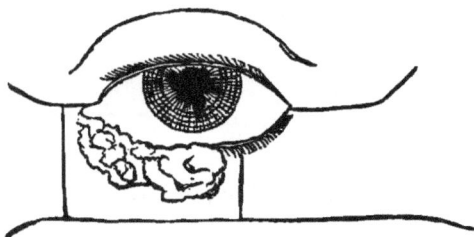

FIG. 25.—Malignant tumor involving the inner two-thirds of the lower lid. Showing the incisions (Knapp's method) made in order to fill the gap resulting
from its removal, by sliding flaps.

as generally the most satisfactory and least disfiguring
one. It consists in the following procedure: If the newformation involves, for instance, the inner two-thirds of
the lower lid-margin and eyelid, we shall have, after its
removal, an extensive gap between the inner canthus and
the remaining healthy part of the eyelid. To fill this

gap, we make an incision through the outer canthus in a horizontal direction towards the temple, allowing its end to run slightly upwards, and a similar incision from the outer lower angle of the loss of substance outwards towards the ear, allowing its end to run slightly downwards. The nearly rectangular flap thus formed contains at its nasal end the remaining healthy portion of the eyelid. When this flap is carefully dissected from the underlying tissues, it is best to try whether, without dangerous stretching, it will cover the gap. This is usually not the case, and another small flap must be dissected from the inner canthus, and from the side of the nose. These flaps are drawn over the gap and are carfully stitched together and to the skin below. Although this newly-formed eyelid presents now a raw wound-surface towards the ocular conjunctiva, it gradually becomes lined during the healing process by an epithelial coat, derived from the part of the conjunctiva, which has been preserved. The disfigurement caused by the scars is trifling, when the wound-lips have united well.

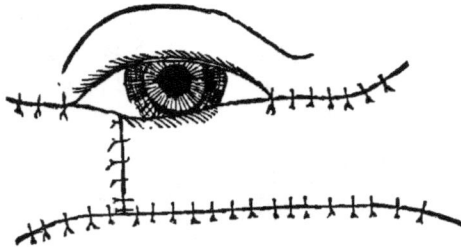

FIG. 26.—Showing the sliding flaps united by sutures.

In some cases of tumor of the lower lid (*Denonvilliers*) it may be well to stitch the margin of the upper lid denuded of of the cilia-bearing border to the wound lips of the gap resulting from the excision of the tumor. When all the parts are well healed together, a new palpebral fissure is made by carefully cutting transversely through the upper lid in front of the cornea. In this way two lids are made of one and the eye can still be opened and closed at will.

In some cases it may be necessary to supply the loss of substance by means of twisted flaps.

Every case, however, must be operated upon according to its own conditions, and the general rules applied to plastic surgery must govern us.

The same, or similar (flap without a pedicle) methods may be applied, whenever a part of the eyelid is destroyed by some other cause. Care must, however, always be taken to preserve whatever is left of a healthy lid-margin and eyelid.

Teleangiectatic and angiomatous growths are not infrequent on the eyelid, especially on the upper ones. They form reddish or dark bluish, soft tumors under the skin of the eyelid sometimes reaching deep into the orbital tissue, and are usually congenital. They are compressible to a certain degree and increase in size when the patient stoops, cries or coughs. These tumors ought to be removed, completely and as early as possible, and the knife or scissors is the preferable means for their removal. Injections of sesquichloride of iron, the use of the actual and galvano-cautery, etc., are less reliable or are followed by a more disfiguring scar.

§30. Different forms of disease of the eyelids and of the palpebral conjunctiva cause the eyelashes to grow in an abnormal direction. This condition is called *distichiasis* or *trichiasis*. It becomes very annoying as soon as the eyelashes touch the eyeball, as the cornea is continually scratched by them.

This constant irritation of the corneal tissue causes it to become inflamed and often to partially or totally lose its transparency by the formation of scars. The trouble is most easily remedied in its incipiency.

It is very common for such patients to pull out the offending cilia, as well as they can, with all sorts of instruments and thus to relieve themselves for a time. The surgeon should, however, not be satisfied with such a palliative remedy, the effect of which vanishes after a few days. A lasting effect can only be produced by a surgical operation, which forces the eyelashes to stand in a direction from the eyeball.

If there is only one or if there are but a few eyelashes which rub against the cornea, they may be removed with their bulbs by the simple excision of a small wedge of tissue from the lid-margin, including their bulbs. Electrolysis which is used by some surgeons for the destruction and removal of lashes, is a very painful procedure, and appears to prove very often a source of future disappointment.

Of the different operations devised for the cure of trichiasis the method of *Hotz* (See Fig. 27) is a very effective one. Its

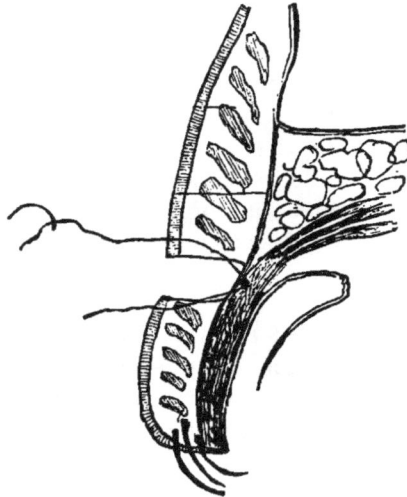

FIG. 27.—Hotz's method of operating for trichiasis. Showing gap resulting from excision of skin and muscle down to the tarso-orbital fascia at upper edge of tarsus and the manner in which the sutures are passed to unite the gap.

main point is that the skin of the lid is forced to adhere to a fixed point, for which he has chosen the tarso-orbital fascia, in the upper eyelid just above the tarsal tissue, in the lower one just below it. An incision is carried through the skin and muscle down to the fascia along the upper edge of the tarsus. A strip of the muscular tissue is then removed and the parts are united by four or five sutures, going first through skin and fascia, and then through fascia and skin on the opposite side of the wound. The simple method yields good and apparently lasting results even in bad cases of trichiasis.

More recently the implantation of a flap of skin between the lips of the lid-margin, which is split in two portions by a vertical incision along the lid, has been successfully practiced.

Trichiasis is very frequently accompanied by a change in the curvature of the tarsus, which causes the edge of the tarsus also to rub against the eyeball. This condition is called *entropium*. It is mostly the result of chronic trachoma of the conjunctiva, commonly called granulated eyelids.

A large number of operations have been devised and are used to remedy this troublesome affection, which materially endangers the usefulness of the eyes. A very successful method is to cut a wedge-shaped piece out of the tarsal tissue, near and parallel to the lid-margin, after having removed the corresponding strips of skin and muscle. When the wound-lips are united by sutures, the lid-margin is turned outwards and thus relief is obtained Relapses are comparatively rare after this operation, which, however, shortens the eyelid [*Snellen-Streatfeild*].

Hotz's operation, just described, does also well in milder cases of entropium. In severer ones I perform *Green's* (See Fig. 28)

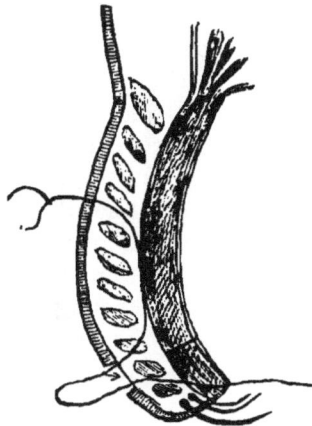

FIG. 28.—Green's method of operating for trichiasis and entropium. Showing incision through tarsus on the inner side and excision of skin on the outer side, and the manner in which the sutures are passed through the tissues before being tied.

operation with preference. The lid-margin in this method is freed by an incision through the conjunctiva and tarsal tissue, running parallel with and about two millimeters removed from it. When by this incision all tough bands of tissue have been severed, the tension is at once relieved. To render this momentary effect permanent it is usually necessary to remove a narrow strip of skin opposite the tarsal incision and to insert a few sutures, which are entered near the posterior edge of the lid-margin, brought out at the lower wound-lip, then entered again at the upper wound-lip and running along for some distance on the tarsal tissue come out through the muscle and skin. These sutures may be removed the next day, or they may be allowed to remain a few days. The requirements in each case may alter the procedure slightly.

In some cases it may be necessary to combine canthotomy or canthoplasty (see later) with the operations for trichiasis and entropium.

§31. When the lid-margin is turned outwards, away from the eye-ball, the condition is called *ectropium*. This concerns mostly the lower eyelids, while entropium is observed on the upper eyelids especially. It very frequently affects chiefly the nasal part of the eyelid, but it may, of course, involve the whole of it. It is usually caused by the shrinkage of cicatrices of injuries or burns, or following inflammatory conditions of the skin, as in blepharitis marginalis.

The epiphora and stillicidium (dropping of the tears), and the irritation and often very great swelling of the conjunctiva, caused by its continued exposure to the air, call for surgical interference, and often the operation required seems very extensive in comparison with the seeming triviality of the affection. This is especially the case in that most frequent form of ectropium, when it is due to the retraction of cicatricial tissue.

If the ectropium is small it may often be remedied by the excision of a rhomboid piece of the whole thickness of the eyelid with the long diagonal in a vertical direction (*Adams*). (See Fig. 29). If the wound-lips are now sewed together, the formerly everted part of the lid-margin will at first be raised considerably above the neighboring parts of the lid-margin. Later on the retraction of the scar will bring it down to the proper level.

In other cases the removal of a triangular piece of tissue from the outer angle of the palpebral fissure and stitching the corner of the lower eyelid into the upper corner of the wound will serve to overcome the eversion.

FIG. 29.—Adams' operation for ectropium.

When the ectropium is, however, very extensive, it will require the removal of the scar-tissue and some form of a plastic operation. In this we may make use of twisted or sliding flaps, or flaps without a pedicle, according to general surgical rules, and as it seems best for the case under consideration.

§32. Drooping of the upper eyelid with a total or partial inability to lift it enough, to expose the pupil for convenient sight, is called *ptosis*. It may be either congenital or due to an acquired paralysis of the levator palpebræ superioris muscle. If congenital, it usually concerns both upper eyelids and the levator palpebræ superioris muscles are atrophic or totâlly wanting. The paralytic ptosis is frequently one-sided.

Von Graefe's method of operating for congenital or paralytic ptosis, consists in the removal of a horizontal fold of the skin and muscle from the upper eyelid followed by stitching of the wound. In doing this the aim must be to shorten the eyelid sufficiently for convenient vision and yet, to leave it long enough to cover and protect the cornea during sleep. In consequence the result of this operation is usually rather inadequate, except in slight cases, and it has fallen somewhat into discredit. *Macnamara* recommends to combine with the excision of the skin and muscle an artificial elongation of the pupil downwards (by iridectomy), and this seems to be a reasonable procedure and calculated to make this form of ptosis-operations more useful.

Two more modern operative procedures, devised respectively by *Pagenstecher* and *Panas*, make use of the frontalis muscle to raise the lid. In *Pagenstecher's* method loops of thread are passed in through the tissue above the eyebrow, and having been brought out on the surface of the skin of the lid are tied on rolls of plaster or glass beads. These loops are allowed to remain until suppuration has taken place in their channels. In this manner cicatrices are formed which are said to allow the action of the frontal muscle to successfully lift the eyelid. *Panas* tries to reach the same result in what seems to be a

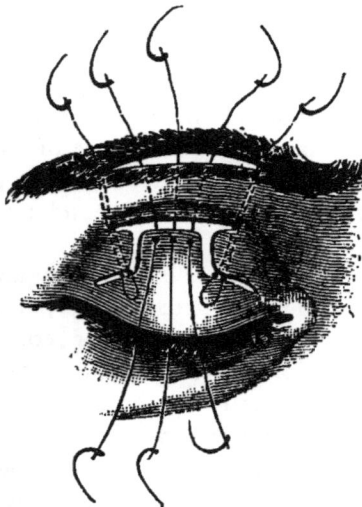

FIG. 30.—Panas' method of operating for ptosis of the upper lid.

more efficient and surely is to the modern surgeon, a more acceptable manner. He forms a quadrangular flap at the upper part of the lid which is freed from the underlying tissue. Then a horizontal incision if made through the skin above the eyebrow. The bridge between the upper end of the flap and this incision, formed by the eyebrow and the skin below it, is now undermined, and the lid-flap is drawn upward underneath it and stitched to the upper wound-lip of the incision above the eyebrow by means of a loop of thread. (See Fig. 30).

Advancement of the tendon of the levator palpebræ superioris muscle has also been tried (*Eversbusch*).

In the paralytic form of ptosis an operation must, of course, not be resorted to, until internal and galvanic treatment have been tried and proved unseccessful. It is most frequently due to syphilis and then usually yields to antisyphilitic treatment.

There is another affection very similar to ptosis, which is chiefly observed in older people and ‘which is not due to a muscular affection but to a superfluity and excessive looseness of the skin of the upper eyelid. It is called prolapse of the skin of the eyelid or *ptosis atonica*. A similar condition is also observed on the lower lid of old people, causing the lashes to irritate the eye. In these affections *Hotz's* method of operating for trichiasis, as described above, is most satisfactory.

§33. The orbicularis palpebrarum muscle is sometimes subject to tonic and clonic spasms, called *blepharospasmus*. In the incipient stage such spasms may concern only a few muscular fibres, and they are then felt and seen, as a slight tremor of the skin of the eyelid near the lid-margin. This slight degree of blepharospasmus is not rare, and it is often observed in overworked individuals, or after excesses in venere or in baccho, and will then disappear without treatment. In other cases it develops into a more serious form, in which the patient is forced to wink his eyelids almost continually, especially when trying to gaze steadily at something, when watched by another person, or, when in the least excited. These spasms may be clonic and tonic at the same time. In such cases the blepharospasmus is, as a rule, combined with similar clonic spasms of the facial muscles. The spasms may be unilateral, but are usually bilateral. This affection is, of course, a most annoying one and it is very difficult to cure. Overstretching of the orbicular muscle, subcutaneous injections of morphine, and local applications of the constant current seem

to yield the best results.   In very severe cases neurectomy of
the supra-orbital and infra-orbital nerves must be tried.

Spasmodic entropium of the eyelids and chiefly of the lower
lid is the consequence of a tonic spasm of the orbicularis palpebrarum muscle, and is frequently observed in affections of
the cornea and conjunctiva, especially in children.   It generally disappears when the irritating cause is removed, but may
sometimes necessitates canthotomy or other operations.   (See
later on).

§34.   *Paralysis of the orbicularis palpebrarum muscle* causes
inability to close the eye or to wink the eyelids, so that the
cornea remains unprotected even during sleep.   This affection
has been called *lagophthalmus* (hare eye), as an old fable states,

FIG. 31.—Method of freshening the lids for tarsorraphy in lagophthalmus.

that the hare sleeps with open eyes. It is one of the symptoms
of paralysis of the facial nerve and may be either a paresis or a
total paralysis.   The dangers arising from the continued exposure of the cornea are evident.   If the paralysis is of long
standing, and no longer curable by the treatment for the nervous disease, these dangers to the eyeball will call for aid from
the ophthalmic surgeon.

The cornea may be partly protected and the eversion of the
lower lachrymal punctum, which always occurs in the later
stages of this affection, may be relieved to a certain extent by
shortening the palpebral fissure.   This little operation, which
is called tarsorraphy, consists in removing a small strip of
skin including the hair-bulbs from the lid-margin of both eyelids to an equal distance from the outer angle of the palpebral
fissure.   (See Fig. 31).

The pared edges are then sewed together.   In some cases

it may be necessary to shorten the palpebral fissure from the inner angle also. and then care must be taken not to interfere with the lachrymal drainage apparatus.

§35. When the palpebral fissure has become shortened in consequence of the shrinkage of the conjunctival sack, and of changes in the curvature of the tarsal tissue, the condition is called *blepharophimosis*. This affection is nearly always due to chronic trachoma.

In order to extend the palpebral fissure, *canthotomy* is performed. This consists either of a single cut through the outer canthus with one clip of a strong pair of scissors, or the cut is followed by stitching the conjunctiva into the corner of the wound (*canthoplasty*). In the former case care must be taken

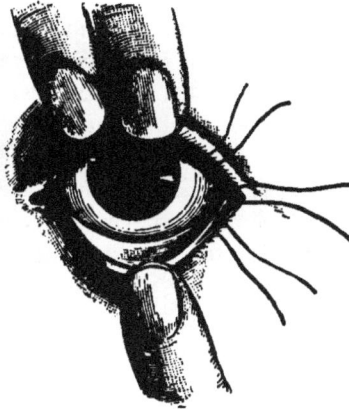

FIG. 32.—Canthoplasty. Showing the sutures by which the conjunctiva is stitched into the gap produced by canthotomy.

that the wound does not heal per primam, or the effect will be lost again altogether. Even, if the healing per primam is successfully prevented, the greater part of the effect of the canthotomy will always be lost. To prevent too much loss, it is best to undermine the conjunctiva near the wound with fine scissors and then to sew it into the gap, caused by the canthotomy. Three sutures, one in the corner, one upwards and one downwards, are usually sufficient. (See Fig. 32).

In severe cases of blepharophimosis *Noyes* has recommended to take a small flap from the temple near the canthotomy-wound and to twist it and sew it into the gap.

§36. *Wounds* of the eyelid, which involve only the skin, or the skin and orbicularis muscle, heal, as a rule, readily when aseptic. Only when they are very extensive, may they give rise to ectropium by the subsequent contraction of the scar-tissue.

When the wound extends through the whole thickness of the eyelid, and reaches the lid margin, it usually remains more or less open, the wound-lips become covered with epithelium, and a traumatic coloboma of the eyelid is the result. Such a coloboma in the upper eyelid may, if extensive, affect the eye injuriously by depriving it of its normal protection, as in the case of paralysis of the orbicularis muscle. If it is situated in the lower eyelid, it forms a very disagreeable ectropium, over which the tears continually trickle down the cheek.

By paring the lips of the coloboma and sewing them together, the deformity may be greatly lessened, and in most cases entirely cured, except for a small notch at the lid-margin.

If a wound severs the fibres of the levator palpebræ superioris muscle, a traumatic ptosis must result. By paring the edge of the torn levator muscle and stitching it to the tarsal tissue the ptosis may be cured.

Other injuries of the eyelids have to be treated according to general surgical principles.

Sometimes we have occasion to see a case of *emphysema* of the eyelids. This is almost always due to an injury with fracture of the inner wall of the orbit (lamina papyracea ossis ethmoidei or os lachrymale). establishing a communication between the nasal cavity and the areolar tissue of the orbit. The patient must be guarded against blowing the nose for a few days. No other treatment, except, possibly, a compressive bandage, will be required.

Hæmorrhages into the tissue of the eyelids, aside from the unsightliness due to them, are of little importance and become absorbed. This absorption may be hastened by mild massage. The common usage of applying leaches to such a suggillated lid is unreasonable, and by causing a superficial wound, may be the means of infection and subsequent suppuration.

Burns of the outer surface of the eyelid are very frequently the cause of ectropium from the retraction of the scar.

Burns of the conjunctival surface of the eyelid and eyeball will mostly result in an attachment between the two. If this is only partial, it is called symblepharon, if total or nearly so, anchyloblepharon. (See Chapter VII). The name of anchyloblepharon is also applied to a union of the lid-margins only.

CHAPTER IV.—DISEASES OF THE LACHRYMAL APPARATUS.

§37. The diseases of the lachrymal apparatus must be divided into diseases of the organ, which secretes the tears, the lachrymal gland, and those of the drainage apparatus, which consists of the lachrymal puncta and canaliculi and the lachrymal sack and duct.

The lachrymal gland in the normal condition secretes continually a small quantity of a clear, alkaline fluid, the tears. An *excessive secretion* of tears is brought about by nearly all irritations and inflammations of the membranes of the eyeball and eyelids, by the presence of a foreign body in the conjunctival sack, and under the influence of emotion, as in crying.

In paralysis of the trigeminus nerve the secretion of tears may be stopped. This will also occur when from some cause the lachrymal gland has become atrophied, or when the tear-ducts have become obliterated.

It happens, although comparatively seldom, that the lachrymal gland becomes the seat of an inflammatory process. This is called *dakryo-adenitis*, and is usually an acute inflammation. It is sometimes seen to attack the lachrymal glands of both eyes at the same time. In some cases dakryo-adenitis is associated with mumps.

In this affection the gland is painful to the touch; it swells and the temporal half of the upper eyelid becomes œdematous. Soon the eyelid swells more, and can no longer be raised sufficiently to open the eye, and the eyeball is pushed gradually downwards and towards the nose, and somewhat out of the orbit. Any movement of the eye in an upward and outward direction is attended with great pain, or such movements may become altogether impossible. It may now be possible to feel by palpation a hard tumor in the outer upper part of the orbit through the upper eyelid, or even to see it protruding into the upper fornix of the conjunctiva, if the swelling permits the upper eyelid to be everted. Soon the tumor becomes softer, and it may then perhaps be possible to detect fluctuation in it. When not interfered with, the abscess may point through the upper eyelid, or through the conjunctiva, and thus be evacuated into the conjunctival sack. The wound heals quickly as a rule, and is but seldom followed by a fistula. In some cases no pus is formed, and the inflammatory symptoms subside gradually and without interference

If the patient is afraid of the knife, all that can be done in this affection is to apply hot fomentations to hasten the formation and evacuation of the pus. The best method of dealing with a dakryo-adenitis, however, is to make an early incision into the swelling, either through the upper eyelid, or, which is preferable, through the fornix of the conjunctiva. This incision should be followed by the application of hot fomentations and the instillation of an antiseptic wash.

A chronic non-suppurative inflammation of the lachrymal gland has been observed, but it is extremely rare.

The lachrymal gland is sometimes the seat of a *neoplasm*. Quite a number of tumors of the lachrymal gland have been described by the older authors, and their histological diagnoses vary greatly, and are little creditable. More recent investigations seem to show that the tumors of the lachrymal gland are usually either of an adenoid or an epitheliomatous character.

In two cases, upon which I had occasion to operate, I found the tumors to be sarcomatous and to consist of short spindle-cells.

By the gradual and mostly painless swelling of the lachrymal gland the eyeball is more and more pushed downwards, towards the nose, and out of the orbit, and the movements of the eye in an upward and outward direction become restricted. As the upper eyelid is usually but little swollen, although apparently stretched and elongated, and the pupil generally remains uncovered, the patient may be greatly troubled by double vision. Soon the tumor may be detected by palpation; its situation in the upper outer part of the orbit and its immobility are sufficient to settle the diagnosis. Sometimes the tumor may be seen in the upper fornix of the conjunctiva. The vision of the eye on the affected side ultimately becomes impaired, and may even be totally lost. This is due to the stretching of and the impeded circulation in the optic nerve, leading to œdema of the optic papilla, optic neuritis and subsequent fatty degeneration and atrophy of the optic nerve.

The tumors of the lachrymal gland seem to develop mostly after injuries. There is nothing to be done in the way of treatment but to remove the tumor. To accomplis this, it is best to make an incision through the upper eyelid, over the seat of the tumor and parallel with the upper orbital margin. When the tumor has been once reached by careful dissection, it should be separated from the sourrounding tissues by means of a blunt instrument, such as the handle of a scalpel, these tumors being, as a rule, soft and easily broken into. It is, of course, best, if possible, to remove the newformation as a whole. The wound usually heals with great rapidity, and the disagreeable symptoms disappear, excepting, of course, such impairment of vision as may be the result of lesions of the optic nerve. Care must be taken not to cut the levator palpebræ superioris muscle, an accident which would lead to ptosis of the upper eyelid.

Some surgeons prefer to attack the tumors of the lachrymal gland through the upper fornix of the conjunctiva, after having first made canthotomy.

The nature of these growths is, it appears, but rarely very malignant, and local relapses are seldom observed, except in sarcoma of the gland.

I have recently had occasion to observe a case of adenoma of the accessory lachrymal gland which did not involve the larger gland although it spread deeply into the orbit. This tumor caused unceasing pain, spastic contraction of the eyelids, some exophthalmus and gradual atrophy of the optic nerve. After two attempts at removal with preservation of the eyeball relapses took place, and finally a thorough removal could only be accomplished after enucleation of the eyeball.

In rare cases a cystic distension of the lachrymal gland or its ducts has been observed; the affection has been called *dakryops.* The cystic character of the tumor may possibly be detected by palpation through the skin. To get rid of it, the whole cyst-wall should be removed, or shrinkage may be induced by injections into the cyst, or by in some manner causing an artificial adhesive inflammation.

§38. We come now to the diseases of the drainage apparatus.

Eversion of the lower punctum lachrymale causes the tears to run down over the cheek. This condition is called *epiphora*, or *stillicidium.* It may be brought on by conjunctivitis or blepharitis; or it may be due to paresis or paralysis of the orbicularis palpebrarum muscle. It accompanies all forms of ectropium of the lower eyelid, and may be due to various other causes.

Simply slitting the canaliculus, the punctum of which is everted, suffices usually to do away with epiphora, when it is not due to paralysis or to excessive ectropium.

Injuries, especially burns, may produce closure of one or both puncta, or the same result may follow chronic inflammation of the lid-margin; cases of congenital obstruction of the puncta are also occasionally met with. Spasmodic contraction of *Horner's* muscle around the base of the lachrymal papilla may constrict a canaliculus. In other cases the presence of a foreign body may cause an obstruction. A small calculus, an eyelash caught in a punctum, threads of *leptothrix* (a vegetable parasite), and various other foreign bodies have been found in such cases. In some cases a small polypus has been found growing from the mucous membrane of the canaliculus and plugging it. The symptoms due to the obstruction of a canaliculus (it is usually a lower one) by a foreign body are slight pain, swelling and epiphora. Slitting the canaliculus and removing the foreign body will give immediate relief. If the punctum is only contracted, it may be sufficient to overstretch it by means of a large pin or small conical probe.

Like the puncta, one or both canaliculi, or the single canal formed after their union before entering the lachrymal sack, may become closed in consequence of injuries or burns. It is then usually impossible to re-establish a communication with the sack.

The mucous membrane of the lachrymal sack and nasal duct is frequently the seat of inflammatory processes. These may come directly from inflammations of the mucous membrane of the nose, or they may originate in the lachrymal sack or nasal duct.

§39. The most frequent form of inflammation of the sack is the catarrhal inflammation, *dakryo-cystitis catarrhalis.* This is attended with hyperæmia, swelling and hyper-secretion of mucus, so that the entrance into the nasal duct becomes blocked up, and the tears cannot pass down into the nose. In this way the lachrymal sack becomes gradually filled with fluid and slightly distended. Lachrymation follows, and the eye seems to be standing in water nearly all the time. Then the lachrymal caruncle and conjunctiva become inflamed, and sometimes blepharitis ciliaris is developed.

The distended lachrymal sack can be seen as a small swelling just below the ligamentum palpebrale internum. If this distension progresses, this swelling may reach a considerable size, and cause the skin above it to become . atrophic. This condition has been called *mucocele.* Even the lachrymal bone may yield to a certain extent to this continued pressure.

By pressing upon the swelling in an upward direction, it is sometimes possible to squeeze the contents of the distended lachrymal sack into the conjunctival sack, and the mucus will then be seen oozing out through one or both of the lachrymal puncta as a viscid, grayish or clear fluid. Or it may be possible to empty the lachrymal sack into the nose by pressing upon it with a downward sweep.

Dakryo-cystitis catarrhalis may go over into a *dakryo-cystitis purulenta,* or the latter form may occur as a primary affection. Both forms of inflammation of the lachrymal sack may, furthermore, be due primarily to inflammation and consequent obstruction of the nasal duct, causing retention, disintergration and infection of the tear-fluid.

In dakryo-cystitis purulenta the contents of the sack are of a yellow, brownish or greenish color, and are often very fœtid. The symptoms noticed in the catarrhal dakryo-cystitis are present in an aggravated form, and there is often considerable pain.

In the severe forms of catarrhal dakryo-cystitis and always in purulent dakryo-cystitis a free opening of one or both canaliculi and sometimes even of the conjunctival wall of the lachrymal sack (*Agnew*) must be made in order to be able to remove its contents. The little operation of slitting the lachrymal canaliculus is best done with *Weber's* curved or straight canaliculus-knife. The blunt end of this instrument must be pushed through the canaliculus into the lachrymal sack until it has reached the nasal wall, before the cut is made. The latter is completed while the knife is withdrawn. If there is no obstruction in the nasal duct, the treatment of the mucous membrane of the lachrymal sack by injections of solutions of bichloride of mercury 1 to 2,500, or weaker, boracic acid 4 to 100, and pyoktanine 1 to 500 will soon restore the normal condition.

It is well, moreover, to direct the patient to use gentle massage over the lachrymal sack, and to direct him to pour a solution of bichloride of mercury into the conjunctival sack, while reclining three or four times a day.

Purulent dakryo-cystitis sometimes gives rise to a phlegmonous inflammation of the subcutaneous tissue in front of the lachrymal sack. This causes a great amount of swelling, heat and redness and often excruciating pain. The eyelids, and the surrounding tissue (nose, forehead and eyelids of the fellow-eye) may become so much swollen, that the condition very much resembles an attack of erysipelas and is not infrequently confounded with it. In these cases it is utterly impossible to squeeze out the contents of the lachrymal sack, even if in spite of the swelling the canaliculi are successfully slit. It therefore may become necessary in such a case to make an incision through the skin and lachrymal sack down to the bone. If this is done at an early stage, great surffering will be prevented. If the abscess is allowed to open by itself, *lachrymal fistula* may be the result. Injections of antiseptic solutions, hot fomentation, gentle massage and the careful removal of all further discharge should follow this incision. Although this mode of operating leaves no disfiguring scar and it is not likely to be followed by the formation of a lachrymal fistula, it is better to empty the lachrymal sack by slitting the canaliculi, whenever this is possible. In a great many cases, how-

ever, the swelling of the mucous membrane will prevent the
emptying even when the canaliculi are slit.

Dakryo-cystitis may, through the formation of scar-tissue, give
rise to the formation of lasting folds or impassible strictures in
the lachrymal sack or nasal duct, which may prevent the
tears from flowing down in the normal way. If a free opening
of the lachrymal canaliculi and treatment by antiseptic and
astringent injections do not bring the mucous membrane back
to its normal condition, we must sometimes be content to advise the the patient to squeeze out the contents of the lachrymal sack as often as it becomes filled, if possible, into the
nose, if not, into the conjunctival sack.

As the pus coming from a suppurating lachrymal sack is
known to be very infectious and to cause ulceration or abscess of the cornea, especially when the epithelium of this
membrane is in the least abraded, the frequent instillation of
one or the other of the antiseptic solutions mentioned above
cannot be too heartily recommended.

Both forms of dakryo-cystitis may appear acute or chronic.

§40. Dakryo-cystitis may be a primary affection, but it is
nearly always caused by *strictures* at the entrance to the nasal
duct or within it, and consequent retention and decomposition
of the tear-fluid. The treatment of the lachrymal sack must,
therefore, in most cases be conjoined with the exploration and
dilatation of such strictures of the duct.

If the symptoms of such a secondary dakryo-cystitis are
not very severe, it will suffice to slit one canaliculus, and preferably the upper one. As soon as this is done, it is well to enter the sack with a probe and explore, whether the entrance into the nasal duct is free. When this is not the case,
it is best to wait a few days, until, under the use of cold compresses and antiseptic solutions, the inflammatory swelling has
subsided. Then a careful exploration of the lachrymal sack
and the nasal duct should be made with a medium sized
probe of the kind devised by *Bowman*, made of silver, or
better, aluminium. I generally use No. 3, although in rare
cases, only No. 2 or even No. 1 can be passed at the
first exploration. There is more danger of making a false

passage with very small probes, and they should therefore be avoided. (See Fig. 33).

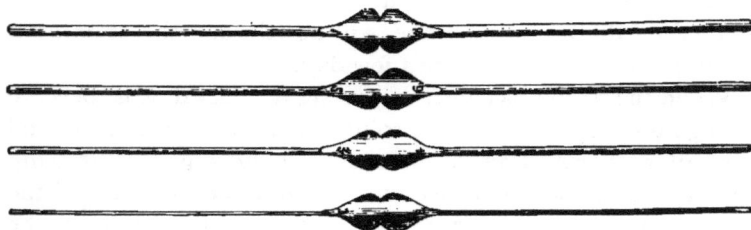

FIG. 33.—Bowman's probes for exploring and dilating strictures of the lachrymal duct.

The probing is done in the following manner: The patient being directed to look downwards the probe is gently pushed through the slit canaliculus, into the lachrymal sack until it reaches the opposite (bony) wall. That the wall is reached, may be judged by the feeling of solid resistance, and by the fact that movements of the probe in a horizontal direction, do not cause the skin of the upper eyelid to become wrinkled. During this first step of the little manoeuvre, it is well to draw the eyelid towards the temple. As soon as it is certain that the probe has reached the opposite wall of the lachrymal sack this traction of the eyelid must be released. Now sliding the point of the probe down along the wall of the sack it is brought into a nearly vertical position, slanting slightly towards the centre of the forehead. It will now be in the position to be slipped down into the nasal duct. This last step is the most difficult, and ought not to be attempted by an inexperienced hand, as a slight pressure in the wrong direction may cause the formation of a false passage, which will not only not accomplish the wished for result but will cause great pain, and sometimes profuse hæmorrhage and may lead to further complications. If the point of the probe has once entered the nasal duct it must be gently pushed down, until it has passed through its nasal orifice. We must make sure of this latter point, as obstructions at the nasal orifice are by no means rare. While the probe is passing down the nasal duct the feeling of resistance will give us an exact idea where a stricture is situated. During this whole procedure no undue force must be used. Often when an obstacle is felt a twist of the probe will help to surmount it. If the lining of the duct is pierced by the probe the bone will be felt, and the attempt at probing must be stopped for a few days.

When the probe has been successfully passed, it is best to syringe one of the antiseptic solutions through the sack and duct after its withdrawal. However, only a gentle pressure must be used, and even this must be stopped at once if the patient complains of severe pain. I have seen effusions of the injected fluid, probably through the walls of the lachrymal sack into the adjacent tissues, which were as alarming in their aspect as they were painful to the patient.

When, by probing, we have located one or more strictures in the lachrymal duct, the further treatment, aside from instillations of an antiseptic solution, consists in their gradual or forcible dilatation.

The gradual dilatation is in most cases the preferable method. After a probe of a certain diameter has been introduced, it is allowed to remain in the duct for ten or fifteen minutes, and then gently withdrawn. When this probe can be easily introduced, a little larger one is used, and again, when this can easily be passed down through the stricture, the next larger one is employed. The probing should be done at first daily, but when large sized probes (No. 5 or 6), pass easily through the duct, the intervals between the probing should be increased. If the patient has pain after the probing, although no false passage has been made, cold compresses or bathing will give relief, and the pain need not interfere with the treatment; yet, it indicates sometimes that too large a probe has been used, and that it would be better to go back to a smaller size. While in children and young people a perfect recovery is generally to be expected, adults and older people are less apt to be perfectly cured, and have generally to be probed again from time to time. In the probing and dilatation of strictures of the nasal duct, I have found the preliminary instillation or injection of a cocaine solution to be of comparatively little value, although I always employ it. When necessary the nasal mucous membrane must also be treated.

Forcible detention by the introduction of a large probe at the first sitting, does not seem to yield as good and lasting results as does the gradual distention; it is, moreover, extremely painful.

When the obstruction of the nasal duct is not due to a stricture in the mucuos membrane, but to an affection of the surrounding bone, which is chiefly caused by syphilis or scrophulosis, successful probing and dilatation of the duct is but rarely possible

If there is a fistula of the lachrymal sack, this will heal without any special treatment, as soon as we have succeeded in restoring the caliber of the duct, so as to allow the tears to flow down through it.

Although it would seem scarcely possible to break off a probe in the nasal duct, I must give warning against this accident, as I once had occasion to remove such a broken probe, left in the duct by the operator.

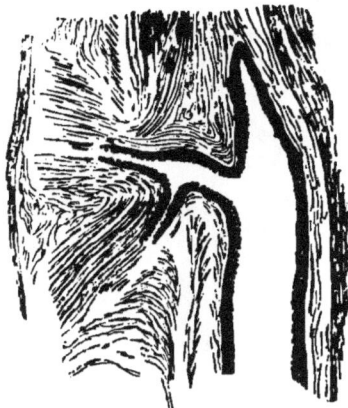

FIG. 34.—(After Gerlach)   Shows the manner in which the two lachrymal canaliculi form one larger canal before entering the lachrymal sack.   The cupola of the lachrymal sack extends considerably higher upward than the point of entrance for this canal is situated.

In a certain percentage of cases the tears will, in spite of successful probing and enlarging of the formerly closed duct, refuse to be drained off in consequence of the lack of elasticity of the walls of the lachrymal sack.   In such cases, and in others also, in which for any reason a re-establishment of the drainage from the lachrymal sack downwards is impossible, the lachrymal sack can be obliterated. This may be done by a free incision into it, followed by the destruction of the mucous membrane by actual or galvano-cautery, or by the use of caustic drugs. Care must always be taken to destroy all of the mucous membrane, and not to omit cauterizing well up in the cupola of the lachrymal sack. (See Fig. 34). The healing by granulation takes place in from two to three weeks. The scar is but slightly disfiguring, and, although the epiphora remains, the patients are greatly benefitted by the operation. The lachrymal sack may also be cut out, instead of destroying its mucous membrane.

Instead of obliterating the lachrymal sack, *Lawrence* advised the removal of the lachrymal gland, an operation which seems to grow in favor.

The lachrymal caruncle is sometimes the seat of a malignant tumor. The most frequent form is *melano-sarcoma*. It must be removed early; the operation is easily performed.

§41. The diseases of the orbit are either diseases of its bony walls and periosteum, or they concern chiefly its contents. In these diseases the eyeball, its muscle, and the optic nerve are as a rule only secondarily affected.

If the disease of the orbit causes the volume of its contents to become increased, the eyeball will be pressed out of its normal position, and, as it cannot escape in any other direction, it will be pushed out of the orbit. This condition is called *exophthalmus*. Exophthalmus may, furthermore, be due to diseases of the neighboring cavities, which encroach upon the orbital cavity, or to tumors of the eyeball itself or of its appendages. The exophthalmus will, in a general sense, be always in a direction opposite to the swelling in the orbit upon which it depends and the movements of the eyeball will be restricted in the direction towards this swelling.

§42. *Periostitis of the walls of the orbit* is usually confined to a part of the orbital margin, and it seems to be more frequently observed in the upper and outer part, than at any other point. In the beginning the patients complain of spontaneous pain, and soon at the seat of this pain a slight immovable swelling appears, which is very sensitive on pressure. Gradually the swelling increases, the conjunctival vessels are hyperaemic, the eyeball is slightly pushed forwards and towards the opposite side, and its movements in the direction of the swelling become restricted and painful. The localized swelling and pain are the most important symptoms for the diagnosis.

In the acute form of periostitis fluctuation will be soon felt, and on incision, or spontaneously, pus will be evacuated.

If the acute form goes over into the chronic form, caries or necrosis of the bone will result, and fistulous openings through the skin will allow of their detection with a probe. The probing ought to be done very carefully, as an acute inflammation of the orbital tissue might follow.

In acute periostitis of the orbital margin, when seen at a very early stage, iced compresses and leeches may sometimes suffice to effect a cure. If not, an incision should be made so as to give a free opening for the escape of the pus.

If the case is already a chronic one, matters will be more complicated. After an incision has been made down to the diseased portion, the necrosed bone must be removed or the carious parts well scraped out. In these chronic cases this operation must often be followed by a plastic one, as the formation of scars bound down to the bone often causes ectropium and lagophthalmus. Orbital periostitis of deeper seated portions may be very difficult to diagnose. Usually the formation of an abscess gives the first clew to the real trouble.

In rare cases a chronic form of the orbital periostitis has been seen to lead to thickening of the bony walls and consequent reduction of the size ofthe orbital cavity.

Orbital periostitis is most frequently the result of an injury. In other cases it is due to a strumous or syphilitic diathesis, and in the treatment we must, of course, take these points into consideration.

§43. Injuries and heavy falls may cause *hæmorrhages* into the orbital tissues. Such hæmorrhages cause sometimes considerable exophthalmus conjoined with paretic symptoms in the muscles of the eyeball. The blood can usually be seen, as it generally infiltrates the ocular conjunctiva, and in some cases raises it so as to form a dark red, shining, ring-shaped elevation around the cornea-scleral margin. Such a hæmorrhage may, of course, be due to the rupture of blood-vessels alone, or it may be complicated by fracture of the bony wall of the orbit. It may, therefore, be a comparatively simple affection, and its alarming symptoms may disappear after a short time without leaving any trace behind. In the other case it may give rise to rather serious complications, and especially, as the fracture often concerns the walls of the optic canal, to atrophy of the optic nerve and consequent blindness.

§44. *Cellulitis orbitæ,* phlegmonous inflammation of the orbital tissue, may be a primary affection, or it may be due to
an injury with or without the subsequent presence of a foreign
body; it may be caused also by orbital caries or necrosis, especially when its seat is far back in the orbit.

Hyperæmia and œdema of the conjunctiva, œdema of the
lids, pain in the orbit, restriction of the movements of the eyeball and exophthalmus are the first symptoms complained of,
and may be attended with fever. The exophthalmus is mostly in a forward direction, and the restriction of the movements
of the eyeball is then general, and not particularly pronounced
in any one direction. The upper eyelid usually swells so considerably that it is impossible for the patient to raise it.

The pressure and traction on the optic nerve may cause
œdema of the optic papilla or optic neuritis, with attendant
amblyopia. Anaemia, followed by atrophy of the optic nerve,
or detachment of the retina, may also be caused by the cellulitis. In other cases sight is apparently not at all impaired.

As the disease progresses pus is formed, and the abscess
may point either in the eyelid or in the conjunctiva, and thus a
spontaneous cure may take place.

In rare cases the pus may break through the lamina papyracea of the os ethmoidei into the nasal cavity, or downwards
into the antrum of Highmore, or even upwards into the cranial
cavity; causing death by purulent meningitis or abscess. The
inflammation may, moreover, extend to the eyeball, and cause
plastic irido-choroiditis, or purulent irido-chroiditis with subsequent shrinking of the eyeball. In other cases the cornea
may slough away in consequence of the impaired nutrition
and exposure due to the exophthalmus.

The formation of pus in the inflamed tissue is occasionally
very slow, and yet, the impairment of sight due to it may be
comparatively small. I once had occasion to see a case of orbital cellulitis five weeks after the first symptoms had set in,
and to evacuate the pus by an incision. The recovery was
rapid, and only a partial atrophy of the optic nerve remained
behind.

In rare cases no pus is formed, and the inflammatory symptoms subside again after a short period of existence. The

manner in which the infection of the orbital tissue comes
about when it is not caused by an injury or caries, or erysipelas is, as yet, unknown. In one case which I have seen lately,
the affection came on during an attack of typhoid fever, and
may have been due to the specific typhoid micro-organism.

With regard to the treatment, first of all, perfect rest and
warm applications should be insisted on. If after a few days,
the symptoms continue unabated, an incision should be made,
whether fluctuation can be felt or not. If pus is found, it will
escape through the incision, and the eyeball will gradually recede to its normal position in the orbit. If no pus is there, or
even if the knife has not reached it, the bleeding and subsequent oozing of serous fluid will reduce the tension of the tissues, and thus give relief. The wound must, of course, be kept
open until all symptoms have disappeared, and the warm applications should also be kept up.

The incision into the orbital tissue is best made through the
conjunctiva. If this is impossible, it may be done through the
lid. A narrow knife should be used and great care be taken
not to wound the eyeball or the optic nerve. It is sometimes
very difficult to reach the pus-cavity, and it is then preferable to
make further exploration with a blunt instrument. The wound
heals very readily, and if the optic nerve and eyeball have remained intact, perfect recovery may be obtained in the course
of a few days. If the incision and evacuation of the pus are
followed by an injection of an antiseptic solution, no undue
pressure must be exerted.

It is well to know that symptoms which resemble very much
those of orbital cellulitis may be due to thrombosis of the
cavernous or of the longitudinal sinus. In the latter cases
however, the affection usually concerns both orbits and their
contents, and, moreover, the cerebral symptoms will help to
make the diagnosis clear.

§45. After tenotomy of one of the external muscles of the
eyeball for strabismus an *inflammation of Tenon's capsule* is
sometimes observed. Although the pain and swelling, and
the visible inflammatory symptoms are somewhat like those of
an orbital cellulitis, a mistake is hardly possible, since the in-

flammation of Tenon's capsule (*Tenonitis,* as it is inappropri-
ately called), causes only a comparatively slight protrusion
of the eyeball.

A small, red swelling is in these cases at first observed near
the incision made in the tenotomy. It is immovable, but may
be fluctuating, and is covered by hyperæmic and œdematous
conjunctiva. This swelling may increase in size and gradually
extend around the whole of the periphery of the cornea. The
movements of the eyeball are painful and somewhat restricted,
but only in consequence of the pain, for the eyeball may be
moved in all directions. The upper eyelid is slightly œdematous. The fluctuation is due to serous fluid within Tenon's
space. This serous form of inflammation of Tenon's capsule
may accompany chronic iridochoroiditis, and it often compli-
cates cases of acute purulent panophthalmitis.

In the uncomplicated cases of serous inflammation of
Tenon's capsule, the fluid may be evacuated by an incision or
aspiration. Otherwise iced compresses, rest and subcutaneous injections of pilocarpine are useful.

Emphysema of the orbital tissue (and eyelids) is sometimes
observed in consequence of a fracture of the lamina papyracea
of the os ethmoidei; also occasionally, in consequence of injuries of the lachrymal bone, or of the bony walls of the nasal
duct from forced dilatation of this passage for the cure of an
obstruction.

§46. The orbit is not infrequently the seat of *neoplasms,*
which may be either benign or malignant in character.

Cystoid formations, when met with in the orbit, usually lie
under the upper eyelid near to and above the outer angle of
the palpebral fissure, and outside of the funnel formed by the
external muscles of the eyeball. These cysts contain an oil-
like fatty, or mucoid fluid, of an amber or brownish tint. Their
walls may become firmly adherent to the periosteum of the
orbit or even to the eyeball itself, and are often very vascular.
They grow slowly and may bring about a partial atrophy of
the eyelid by their continued pressure. They, moreover, displace the eyeball and may cause a noticable exophthalmus
with attendant double-vision, and with the other symptoms

which depend upon continued stretching and impaired nutrition of the optic nerve.

True dermoid cysts, and echinococcus cysts have also been
observed in this locality.

Such cysts must be removed, and, of course, if possible
enucleated in toto. In doing so care must be taken not to
breack the cyst-sack, or to injure the eyeball or its external
muscles.

Of vascular tumors both angiomata and teleangiectatic
growths are found in the orbit.

A number of primary *epithelial cancers* of the orbital tissue
have been described, yet it seems that the tumors of this tissue
are, as a rule, not of an epithelial, but on the contrary of the
connective-tissue type. Thus we find *round and spindle-cell
sarcoma, melano-sarcoma, fibro-sarcoma, myxo-sarcoma, cysto-
sarcoma and cylindroma* of the orbital tissue.

I had once occasion to examine a large orbital tumor, removed fram a negro woman, with preservation of the eyeball,
by the late *Dr. Darby*, of New York. It proved to be a *leiomy-
oma*, and consisted almost wholly of organic muscular fibres.
No similar case has been anywhere reported, and I suppose
the tumor originated from the organic muscular fibres lying
in the orbital tissue.

The symptoms of all these different forms of orbital tumors
are similar, and the most prominent one is always the exophthalmus. To this may be added intercurrent inflammatory
symptoms, and, again, all the symptoms referable to the optic
nerve and to the cornea, as we observe them in cases of exophthalmus due to a newformation in the lachrymal gland or to
orbital cellulitis.

The tumor may frequently be detected by palpation and
does not, of course, move when the eyeball is moved.

Such a newformation must be removed, as soon as detected.
As long as it can be done, an attempt should be made to preserve the eyeball. If a complete removal of the tumor cannot
be accomplished without it, the eyeball must be sacrificed, and
in some cases the whole of the orbital tissue will have to be
cleaned out in order to save the patient's life.

In case the eyeball has been removed with the tumor, a suggestion made by *Green* is very valuable, indeed. It is, to further remove the lid-margin, the palpebral conjunctiva and the tarsal tissue, and sew the eyelids together. The eyelids heal promptly together and thus form a permanent cover, which protects the deeper tissues from injurious influences.

Another class of orbital tumors spring from its bony walls in the forms of *osteoma or periosteal sarcoma.*

The contents of the orbit may, moreover, be invaded by tumors originating in the neighboring cavities, especially the nasal cavity and the antrum of Highmore; the symptoms will be much the same as in cases of primary orbital tumors.

Pulsating exophthalmus, as well as exophthalmic goitre (*Basedow's, Graves' disease*) will be spoken of in Chapter XXIII.

§47.   Continued *cold applications* to an eye are best made by
means of a piece of light linen, which after having been cooled
either in cold water, or ice water, or directly on ice, is laid
upon the closed eye. This linen should be folded several
times, because it will then keep its temperature for a longer
time, and a second one ought always be kept cooling while the
first one is lying on the eye.   As soon as the linen on the eye
no longer causes a cooling sensation, it should be changed.
The time in which this will have to be done, will, of course,
depend on various circumstances.   With children it may be
necessary to fasten the linen with a simple bandage.

Care must always be taken to wring the linen dry before applying it, and not to allow any cold water to trickle down and
enter the ear.   This may be the better guarded against by
putting some oiled cotton into the ears.

Instead of the linen a very small ice-bag may sometimes be
used.   Yet its weight must be so small as not to be felt disagreeably.   A continued current of cold water through tubing
forming a coil over the eye may also be employed.

When cold bathing only is required from time to time, this
is best done by gently pressing a cooled sponge or linen rag
against the closed eye.   This may, however be replaced with
great advantage by an eye-douche.   A quart or more of cold
water may thus be allowed to flow against the eye from a moderate hight through a very fine rose.

Opening the eyes under water should be avoided.

In cases of phyctænular keratitis, with great dread of light
and spasmodic entropium, it is of great value to apply a sudden cold bath to the whole face.   This is best done by plung-
ing the child's face into a basin of cold water and by holding
it there until it struggles for breath.

This procedure may be varied by holding the child over a

basin and directing a moderately strong stream of cold water full in his face through a rather course rose.

Interrupted hot applications are best made in the same way as cold ones, or by allowing steam from an atomizer to be thrown against the closed lids.

Dry heat may sometimes be indicated, and may be best applied by means of the Japanese hot-box or in default of this by means of little bags containing a light material, such as bran which will retain the heat for some time. •

Leeches must never be applied either to the eyeball itself or to the eyelids. The best place for the application of leeches in eye-affections is the temple, or, more accurately, the space between the outer angle of the palpebral fissue and the line, where the hair begin to grow.

This is also the best place for the application of *Hœurteloup's* artificial leech.

Any *discharge* from the palpebral conjunctiva should be gently and carefully removed. What sticks to the eye-lashes and the lachrymal caruncle, when semi-fluid, can be easily wiped away by the use of a soft, moist sponge or absorbent cotton.

When the discharge is dried up and hard, it must first be well soaked by bathing with warm water. It can then be removed by brushing the eye-lashes back and forth in a horizontal direction with a sponge, or still better with a dry towel. After the eyelashes and lid-margins have thus been cleansed, the conjunctival sack must be inspected and every film of coagulated discharge be removed. This may be done by gently wiping it off with a very warm sponge or some absorbent cotton, or better yet by a gentle stream of a 4% boracix acid solution, which is allowed to flow over the eye.

How to evert the eyelids for examination of the palpebral conjunctiva and fornix, has been described in Chapter II.

§48. For the *instillation* of medicated fluids into the conjunctival sack, it is best to use a dropping tube, or dropping glass. Where this cannot be procured a clean teaspoon may answer the purpose.

As infection may undoubtedly be carried from one patient to the other by means of a dropping tube, every care should be had to use it only when in an aseptic condition. In hospital-practice every patient should have his own dropping tube or glass. In order to instill the fluid into the conjunctival sack, the eyelids should be held apart with the other hand and the lower eyelid be drawn down sufficiently to allow the drop to enter, while the patient is directed to look upward. Medicated fluids, which have no astringent or caustic quality, and whose effect is to be reached by absorption, such as solutions of a mydriatic or miotic drug, should, if possible, be allowed to drop directly upon the cornea, as they are thus absorbed more readily. The drop must not fall from any appreciable hight, should not be cold, and must not be allowed to be washed out at once by the tears. Whenever, therefore, it is difficult to make such instillations or when instillations are to be made by the inexpert, it is best to direct the patient to lie on his back and to hold his eyelids apart for some time after the instillation has been made.

Before applying any astringent or caustic remedy to the conjunctiva, the rule should be to instill a few drops of a cocaine solution into the conjunctival sack. By this means the patient will be saved a considerable amount of suffering, and we are much better able to make our application for the very reason that the patient does not dread it.

Astringent solutions should be applied by means of a moderately large camel's hair brush, dipped into the fluid and then drawn across the conjunctival surface of the everted eyelid. It usually suffices to apply them to the lower eyelid. If, in spite of the cocaine, pain and irritation are very annoying after such an application, the patient should be directed to bathe the closed eyes for some time with cold water.

After having been used the brushes must be washed and kept in a solution of bichloride of mercury (1 to 5,000) for hours, or best until shortly before they are used again the next day.

The application of *caustic solutions* should always be made by the surgeon himself. When making such an application the cornea must never be left unguarded. It is therefore best to treat each diseased eyelid separately with a brush. While the caustic solution is brushed upon the inner surface of the everted upper eyelid, the patient looking down, the lower eyelid is gently drawn upward, so as to cover the cornea perfectly. By a similar manœuvre the upper eyelid is made to protect the cornea while the lower one is treated. As soon as the application is made, the brush should be dipped into a bowl of water, held by the patient under his chin, and the superfluous caustic be washed off with it. It is, of course, impossible to neutralize the primary effect of the caustic by washing, so there need be no hesitation in using plenty of water. The application of caustic fluids should not be repeated until the superficial eschara caused by the last application is cast off, which takes place in from 20 to 24 hours. It is therefore best to use these remedies only once in 24 hours and to make the application at about the same hour every day, preferably in the fore-part of the day or at least not in the evening.

By quick, intelligent manipulation with the brush, the effect of the caustic application can be nicely graded, and even to a certain extent localized.

Sulphate of copper, in substance, the sovereign remedy in trachoma, should also be applied by the surgeon himself. For this purpose a large crystal should be trimmed thin and perfectly smooth and mounted in a crayon holder to make its handling easier. No sharp edge or roughness must be allowed to remain before the application, and the surgeon should therefore examine the crystal every time before using it, and dry it well after the treatment of every eyelid. The application to the conjunctiva should be made very gently by simply drawing the crystal once across the part to be treated, and the conjunctiva should then be washed off at once with a brush dipped in water. All rubbing and prolonged contact, producing a caustic effect, is to be avoided. What is wished for is only irritation and stimulation, and not the "burning off" of the granules. It is, therefore, a good rule, to aim at touching the diseased parts as lightly as possible, rather than the reverse. As the seat of the granules is mostly in the fornix, the surgeon must be careful to apply the copper crayon to this part especially. In treating the fornix of the upper eyelid, it is necessary, therefore, to go high up under the everted eyelid, and in doing so it is best to lay the crystal against the lower eyelid, so as to drag it

along and protect the cornea, while the copper is shifted upwards. Treating the fornix of the lower eyelid is, of course,
very much easier, since it can be fully exposed. All these
little manipulations must be performed with great delicacy
and care, in order to obtain the best possible result. In this
point lies the secret of the different results obtained by different physiciaus with the sulphate of copper in substance.

Nitrate of silver in substance or in the form of the mitigated
stick is best altogether avoided.

Some remedies are best applied to the eye in the form of
ointments. The common advice to simply smear a little of the
ointment into the eye or on the inner surface of the lower eyelid, is not sufficient, as only very little of the ointment will
reach the part it is chiefly intended for, namely, the cornea.

The best method of applying an ointment, if it is not fluid
enough to be brushed upon the inner surface of the upper eyelid, is to take a little on the end of a blunt probe, to bring it
between the separated eyelids, and then to close the lids quickly while the carrier is withdrawn. If this does not succeed
the upper eyelid must be everted and the probe with the ointment be brought upwards between it and the eyeball. Then
the probe is withdrawn, and the eyelid returning to its normal
position wipes off the ointment.

In the cocainized, eye it is possible to directly apply the
ointment to the cornea, as no reflex-spasm takes place.

It is then well to gently rub the ointment with the eyelids
all over the eyeball in circular and radiating movements, exerting all the time a slight pressure. These movements have
the effect and value of massage, and are especially to be recommended in corneal affections.

For the *inspergation* of medicinal *powders,* such as calomel or
iodoform, into the conjunctival sack, it is best to make use of
a small, dry camel's hair brush. This is lightly dipped into
the powder, the eyelids are separated with the fingers of the
other hand, and the powder is snapped off the brush into the
conjunctival sack.

As the discharge in most forms of inflammation of the conjunctiva is contagious, a great many efforts have been made to
perfectly *isolate* a healthy eye, so long as its fellow continues

diseased. Yet most of these contrivances are very annoying
and after all useless, as perfect isolation is almost impossible.
Absolute cleanliness and the free use of antiseptic solutions,
particularly to the as yet unaffected eye, according to my experience, are at least as good preventives as any appliance for
isolating a healthy ese, especially when the patient remains in
such a position that no discharge from the diseased eye can
run across the bridge of the nose into the healthy one.

To isolate, however, a child suffering from one of the serious
forms of conjunctivitis (purulent, gonorrhœal, diphtheritic or
trachomatous conjunctivitis) from other children, is highly to
be recommended, nor should such a child ever be allowed to
go to school. This is often allowed when children suffer from
trachoma or from chronic purulent conjunctivitis, but it is to
be absolutely condemned as bad practice.

§49. *Wild hairs*, ingrowing eye-lashes, are easily pulled out,
when they are large and well pigmented, with appropriate forceps. But often the most annoying ones are very thin and fine,
and almost unpigmented. To detect these is then sometimes
rather difficult, and is best done by placing the patient side-
ways to the light, and then looking along the lid-margin,
which should be slightly drawn away from the eye. Alternately applying the lid-margin to the eyeball and lifting it off
will help greatly in the detection of such eye-lashes, since they
will raise the tear-fluid somewhat before being drawn away by
the eversion of the lid-margin. Great care must be taken to
extract the cilia with the root and not to brake them off.

The whole procedure is, however, of but little value, and, to
afford a temporary relief only, must be repeated again and again.
The patients should therefore be persuaded to have an operation for trichiasis performed. (See Chapter III).

The removal of *small foreign bodies* from the conjunctival
sack is, as a rule, a very simple affair, as the foreign bodies lie
for the most part loosely on the conjunctiva. They may be
removed from the everted eyelid by means of a moist camel's
hair brush or a piece of soft linen. In the lower conjunctival
sack they are usually found lying in the fornix. In the upper
conjunctival sack they lie most frequently in the small depres-

sion just above the inner edge of the lid-margin. When one small foreign body has been thus removed, the surgeon should scan the whole conjunctival sack once more, and, if necessary, sweep the upper conjunctival sack as high up as possible with a camel's hair brush, moistened with a boracic acid or sublimate solution. When the foreign body is not easily detected by its color, it may be necessary to draw the aseptic finger gently over the surface of the palpebral conjunctiva and thus to satisfy ourselves of its seat and perfect removal.

For a careful inspection of the eyes of children, it is often necessary, in spite of the previous instillation of cocaine, to separate the eyelids by means of *Desmarres'* or some other kind of *lid retractors.* During their use great care must be taken not to exert, nor to let the patient exert, any pressure on the eyeball. It is therefore best to insert the retractor for the upper eyelid first. This is done by everting the lid-margin slightly by dragging the skin upwards with the index finger of one hand and slipping the retractor under the eyelid with the other hand, all the time pulling the eyelid slightly away from the eyeball. The insertion of the retractor for the lower eyelid is much easier, but must be done in the same manner.

To insert and remove a *wire-speculum* to hold the eyelids apart during an operation on the eyeball requires the same delicacy. The branch for the upper eyelid in most cases should be inserted first, while the branches are held tightly together, and the same precautions are to be observed as in inserting the lid-retractor. While this is being done the patient is directed to look downwards. Then, while the patient looks upwards, the branch for the lower eyelid is inserted, and the two branches are allowed to separate. When the speculum is removed, the branch for the lower eyelid is first taken out, while the branches are pressed together, and then the one for the upper eyelid. In doing this the eyelids are, with the speculum, gently pulled away from the eye-ball, so as to avoid any pressure upon it.

The insertion of a wire-speculum is, of course, necessary, or at least of great advantage, in all important operations on the eye-ball, yet the general practitioner should also be familiar with it, as he is frequently called upon to remove small foreign bodies from the cornea. Some patients, in spite of cocaine instillations, cannot keep their eyes open voluntarily and the lids must be forcibly held open in order to be able to accomplish the desired end.

After the speculum has been properly adjusted, the conjunctiva and subjacent tissue are grasped with the fixation-forceps near the cornea, and best near the lower corneo-scleral margin, as the eyeball will instinctively fly upwards to avoid the instrument used for the removal of the foreign body. The latter should not be a sharp instrument, especially in the hands of the unpracticed operator. A bent needle, or a somewhat blunt minute spud or gouge, are the most appropriate instruments for this purpose.

Care must be taken not to injure the neighboring parts of the corneal tissue, and to attack, as far as possible, the foreign body only. The unexperienced operator, however, had better keep his hands from this apparently trifling, but sometimes rather difficult, little operation.

When inserting an *artificial eye*, it is best to follow the rule for inserting the speculum, that is to insert it first under the upper eyelid while the patient looks downwards, and again, when removing it, to lift it first out of the lower conjunctival sack while the patient looks upwards. The patient should be directed to bend his head over a bed or pillow, when removing the artificial eye himself, until he has acquired sufficient skill in the manipulation to accomplish it without fear of letting the artificial eye fall.

§50. For closing the eye or both eyes after an operation or injury, it has been and still is the custom with a number of oculists to use a bandage of flannel, linen, gauze, or some such material. Such a bandage is from two to three yards long, and from one and a half to three inches wide. After the eye has been nicely padded with some light and elastic aseptic material, the end of the rolled up bandage is placed over the ear on the affected side in a horizontal position, and the bandage unrolled toward the forehead and wound around it. After it has passed above the ear on the opposite side the roll of bandage is lowered so as to reach the face again from be·

low the ear on the affected side, and again unrolled upwards across the affected eye towards the other side of the head, where it may be fastened with pins or tied with braid sewed to the end, or several more such tours may be made. In this manner varying degrees of pressure may be applied to the eye. In the same way both eyes may be bandaged, and to afford more security against the slipping of the bandage a cross-piece may be added going over the top of the head. To prevent the possibility of any unforseen pressure little wire masks (*Prout*) may be applied to the eye. (See Fig. 35).

FIG. 35.—Wire Mask.

The advantage of such or similar bandages have been decidedly overestimated. In spite of all care, they slip often and cause themselves the pressure upon the affected eye, for the prevention of which they have been applied. They are in the way of free movements of the face and head, just on that very account. In warm weather they are hot and fret the patient. When they are not perfectly aseptic they may form a source of infection and, however well applied, they will admit air, which means increased possibility of infection.

A neatly applied adhesive plaster is free from all these objections and is preferable.

The *ideal* closing of an eye, however, according to my own experience, consists of a pad of absorbent cotton moistened with an antiseptic solution, which is shaped so as to fill out the orbital depression near the nose and to cover the whole of the eyelids, and is held in place by a piece of adhesive plaster reaching from the cheek to the forehead, and from the nose to the temple. This plaster can be applied in such a manner that it does not press on the eye, not even during mastication, and makes an air-tight closure. If the dryness of the cotton on the second or third day is felt disagreeably by the patient, a little moistening of a narrow strip allows of the insertion of a dropping tube to re-moisten the cotton with an antiseptic solution. Such a plaster allows of all reasonable freedom of motion, remains aseptic and is a boon to the patient. The adhesive quality must, of course, be of the best, so that even the perspiration will not loosen it.

§51. This is probably the best place to give a few hints for the use of general practitioners who are asked to *assist* in an eye-operation. The tenderness and delicacy of movements required in the operator should also be possessed by the assistant. The observance of a few simple special rules by the assistant will make his services valuable, while no assistant would be preferable to one, not observing these rules.

As good light is absolutely and especially required in every eye-operation, the first rule for the assistant is, to keep out of the light, and, as perfect freedom of movement is necessary for the operator, the second, and no less important rule, is to keep out of his way. Both rules are usually best followed, if the assistant stands on the side of the patient opposite to the operator. As the operator will, as a rule, be ambidexter, the assistant ought to be so too. If, for instance, the assistant is required to hold the eye steady during a certain stage of an operation, he should hold it with the right hand when the operator uses his left, and *vice versa*, at least when they stand on opposite sides of the patient. To have the eye steadied by a trusty hand is of such great assistance to the operator, that he would not be likely to do without it, when it is to be had. Yet, a hand that is not trusty, is worse than none.

In order to steady the eyeball the conjunctiva is grasped with the fixation forceps near the cornea-scleral margin, and as di-, rectly as possible at right angles to the direction of an incision, and the teeth of the forceps are inserted into the tissue as deep as possible. The forceps should always be held so that the thumb can at any moment press on its spring-catch and open it without any further movement. It is best to use forceps without a catch. No traction and no pressure must be exerted. If the globe must be turned downwards by the assistant this is not to be done by pulling downwards on the fixation forceps, but by rotating the eyeball around its horizontal axis, by slightly raising the hand that holds the forceps and lowering the part attached to the globe gently towards the lower fornix. This little manipulation, if awkwardly done, may ruin an eye.

It is sometimes necessary for the removal of a small foreign body from the cornea, or the division of the lens-capsule, etc., to have the field of operation well illuminated by artificial light. To do this we use a large magnifying lens, which, of course must be held by the assistant. It seems an easy matter to throw sufficient light with such a lens upon the field of operation, yet, it requires careful attention on the part of the assistant to do it satisfactorily, as the position of the lens must be changed with almost every movement of the eyeball or head of the patient. The assistant should therefore not divide his attention between this and anything else of interest during the operation, or it may happen that the operator will find himself suddenly with a dark field of operation before him.

When the operation is a bloody one, as, for instance, an operation on the eyelids, or a strabisum operation, or the enucleation of an eyeball, the assistant should be prompt in wiping away the blood. A careful assissant will, by wiping quickly after every cut that draws blood, enable the operator to work rapidly and never in the dark.

It is always best to wipe the blood away, not to soak it up, as is frequently done, by pressing a sponge or absorbent cotton on the bleeding surface. On the contrary all pressure should be carefully avoided, especially in operations upon the eyeball itself.

Lastly, the assistants hands should should be made aseptic before the operation is begun.

CHAPTER VII.—DISEASES OF THE CONJUNCTIVA.

§52. *Hyperæmia* of the conjunctival blood-vessels is frequently observed, especially in the conjunctiva of the eyelids. It may be more pronounced in one part than in another, but it always shows least in the fornix. The color of the hyperæmic parts is a bright red, almost scarlet. In hyperæmia of the eyeball we have seen (see Chapter II), that it is necessary to distinguish between hyperæmia of the ocular conjunctiva, and that which lies more deeply and has its seat in the sclerotic.

When hyperæmia of the conjunctiva has existed for some time, the fornix, and later on the ocular conjunctiva, show a slightly œdematous condition. Later the papillæ of the conjunctiva become enlarged, and protrude slightly above the general surface of the conjunctiva, especially near the fornix.

Hyperæmia of the conjunctiva may be due to some irritation as for instance, a lack of moisture in the air of a heated room, electric light, or too much glaring light from any source; to the presence of small foreign bodies in the conjunctival sack; it accompanies coryza, and it may even originate in the strain incident to an error of refraction. It can, furthermore, be symptomatic both in certain more serious forms of eye-diseases, and in other disorders.

Patients suffering from hyperæmia of the conjunctiva complain of a dry, heated feeling, especially in the evening. Their eyes get easily tired, and there is slight photophobia and lachrymation.

The removal of the irritating cause, which, however, it is sometimes, quite difficult to find, will be followed by the disappearance of the hyperæmia. If there is an error of refraction or accommodation, its correction by suitable glasses will have the desired effect.

When the hyperæmic condition of the conjunctiva has be- .

come chronic, the instillation of 4% boracic acid solution, or even a 1-2% solution of sulphate of zinc will be useful. One of these remedies may be combined with systematic cold bathing or the cold eye-douche. Some patients prefer hot bathing.

A special form of very marked hyperæmia of the conjunctiva, which is chronic and recurrent, is due to syphilis. This will yield only to antisyphilitic treatment. Local applications in these cases are not only useless but they do not even seem to be agreeable.

§53. When hyperæmia of the conjunctival blood-vessels is combined with an increased and abnormal secretion of the conjunctival mucous membrane, we have to deal with *conjunctivitis*.

The protection afforded by the eyelashes and the reflex movements of the lids do not suffice to shield the moist surfaces of the cornea and conjunctiva from the adhesion of numerous micro-organisms which floating in the air come continually in contact with the open eye. It is, therefore, not astonishing that the normal conjunctival sack even contains a large variety of micro-organisms which are of a seemingly non-pathogenic character. Although it not as yet proven for some, it is most likely that all the different forms of conjunctivitis are due to the presence and products of some kind of micro-organism. This explains the fact long well known, namely, that the discharges from the conjunctiva are contagious, that is that any form of conjunctivitis may be transmitted by the discharge, if it reaches another eye in a moist state, and perhaps also when dry. The degree of inflammation produced in an eye by such an infection may vary considerably, according to its susceptibility. Other conjunctivæ may be altogether immune against the special infection. The epidemic and endemic appearance of some of the forms of conjunctivitis is now better understood.

The discharge from an eye suffering from any form of conjunctivitis being contagious, the physician must never forget to guard the family of, or attendants on a patient against conveying any of the discharge to their own eyes by means of the fingers, towels, handkerchiefs, etc. Constant and careful

removal and disinfection, or destruction by fire, of any discharge, and absolute cleanliness and antisepsis must be insisted upon.

It is best, therefore, in all forms of conjunctivitis to give the
patients an antiseptic solution with which to flush the conjunctival sack three or four times during the day, aside from the
local treatment applied by the surgeon himself.

Eyes suffering from any form of conjunctival inflammation
must never be bandaged.

Acute catarrhal conjunctivitis, is characterized by hyperæmia and œdematous swelling first of the palpebral and later
of the ocular conjunctiva, accompanied sometimes by small
subconjunctival ecchymoses. Gradually the papillæ of the
conjunctiva become enlarged and give it an uneven, even velvety appearance. The eyelids, especially the upper one, swell
so, that the papebral fissure appears smaller. Photophobia
and lachrymation are but seldom absent and pain or great discomfort is a prominent symptom. The patients often locate
the latter in the outer or inner angle of the palpebral fissure
and insist upon it that they have a forgein body in the conjunctival sack. These symptoms are accompanied or soon
followed by the secretion of mucus or muco-pus. This is not
secreted in large quantity, and coagulates easily. It will be
found in yellowish flocks within the conjunctival sack, especially in the lower and upper fornix and upon the lachrymal
caruncle. Some of it usually adheres to the eyelashes.
During sleep the eyelashes of the opposite eyelids by the
dried up mucus are glued together and the patient is unable
to open his eyes on waking. This often gives rise to excoriations along the lid-margins. At the outer angle of the palpebral fissure the skin is often found to be red and painful, and
easily torn, which is due to maceration by the tears.

In some cases this complex of symptoms comes on very
violently and is attended with so much oedema and discharge,
that the appearance resembles an attack of acute purulent infection. This form is often endemic or even epidemic, and has
been more frequent than ever since the epidemic of influenza
has made its appearance.

The real living contagium of catarrhal conjunctivitis has not
yet been found.

Although a "cold" has probably little to do with the causation of an attack of conjunctivitis, it is usually attributed to it. The possibility, that germs that live in the conjunctival sack without being able to do harm as long as it is normal, may, when direct chilling of the lids and conjunctiva takes place, become enabled to do harm, cannot be denied. We have analogues to such an occurrence, for instance, in the production of pneumonia by the pneumococci which have lived in the air-passages without doing harm, until they are given a chance, so to speak, "by a cold."

Catarrhal conjunctivitis may give rise to the formation of small ulcers of cornea at its conjunctival margin. When some of these coalesce, a larger crescentic ulcer may result. They disappear, as the catarrhal inflammation of the conjunctiva disappears and usually need no special attention although they are sometimes quite painful. If necessary atropine or eserine may be instilled.

Aside from a frequent washing of the conjunctival sack with an antiseptic solution, the application of a nitrate of silver solution (1 to 100) by the surgeon is the remedy on which most reliance can be placed to cure this disease. This application should be made once in every twenty-four hours. As soon as the nitrate of silver touches the conjunctiva, it gives rise to the formation of a superficial eschara, which is followed by an increase of all symptoms for a period of hours varied according to the individuality of the case and the degree of thoroughness of the application. Cocaine may reduce the pain, but not prevent it. Gradually the irritation decreases and the cast-off eschara may be found lying in the folds of the conjunctival sack as a mucoid string, and be removed. Regeneration of the epithelium now takes place gradually and is finished in about twenty-four hours after an application of the nitrate of silver solution. It is then, that a new application should be made.

This treatment may be aided in and the patient's comfort be greatly increased at first by the continued, later on by the interrupted use of cold cloths or iced compresses to the eye. The photophobia is sometimes so disagreeable that it is best to order the patient to wear smoked glasses. To prevent the lashes from sticking together white vaseline should be applied to the edges of the lids at night.

Acute catarrhal conjunctivitis may pass over into the *chronic* form.   In the latter the symptoms are all considerably ameliorated, and the patients may be able to attend to their duty all day while in the evening they are particularly annoyed by the affection.

The eyes become easily tired and "sleepy" and complaint is often made of a feeling, as if dust or sand were lying in the conjunctival sack.

Another sensation, as if the lids were stuck to the eyes and could not be opened, is also a common complaint.

During sleep some discharge collects at the inner angle of the palpebral fissure or adheres to the eyelashes, and often the eyelashes are found stuck together in the morning.

Like chronic pharyngitis on rhinitis, chronic catarrhal conjunctivitis is one of the most frequent affections, and while some patients will never mind it, others will try all means to get rid of it.   Relapses are extremely frequent, still beyond the annoyance caused by it, the disease, but seldom gives rise to any more serious sequels. In rare cases corneal ulcers may be seen due to it.   When it is combined with blepharitis ciliaris, as it mostly is, it aids this disease in the production of ectropium.

Chronic catarrhal conjunctivitis does not always start with an attack of acute catarrhal conjunctivitis.   It may come on insidiously and is then due to the frequent irritations of the conjunctiva by means of bad or dry air, heat, dust, chemical vapors and similar admixtures to the air.   It may also proceed from a stoppage of the tear passage (*lachrymal conjunctivitis*) or from the presence of a tumor in the lid or conjunctiva, like pterygium, and in a great many cases from strain due to an error of refraction or accommodation.

In chronic catarrhal conjunctivitis the use of a nitrate of silver solution is but rarely indicated.   The affection is reduced in its more disagreeable symptoms or cured by the application of milder astringents.   A one per cent. solution of sulphate of zinc is the best remedy in the average case.   This may be brushed over the conjunctiva once or twice a day.   In mild cases the washing of the conjunctival sack with a four

per cent. solution of boracic acid may be sufficient. To
these remedies may be added cold or hot bathing, according
to the patient's feeling. Other remedies recommended are
alum, tannine and tincture of opium. To prevent the lashes
from becoming glued together during sleep, the edges of the
lids should be greased with white vaseline.

§54. *Acute purulent conjunctivitis (blennorrhœa of the conjunctiva, blennorrhœal or gonorrhœal conjunctivitis)*, is probably always due to the infection of the conjunctiva with the
gonococcus of *Neisser*. These micro-organisms are found in
the epithelial cells and in the discharge and therefore get into
an eye either by direct contact, or indirectly through the air,
or by means of towels from another purulent eye, or from a
patient suffering from gonorrhœa.

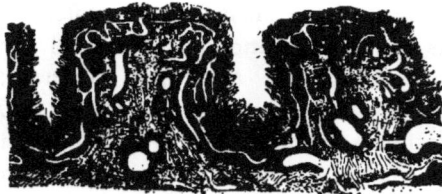

FIG. 36.—(After Saemisch). Hypertrophied papillæ of the conjunctiva in blennorrhœal conjunctivitis.

After the infection has taken place and a varying period of
incubation has passed, the eyelids become hot and swollen.
This œdema of the lids may very rapidly become so considerable that the patient cannot open the eye. At the same time
a watery discharge, intermixed with flocks of yellow muco-
pus, makes its appearance and glues the lashes together. As
these symptoms increase in severity, the conjunctiva of the
lids swells more and more, so that when the lids are everted
the enlarged and engorged papillae form several ridges, with
deep fissures between them. (See Fig. 36). The feelings of
heat, weight and pain grow very distressing. Gradually the
discharge changes to a thick cream-like yellow, sometimes
greenish pus, which continually pours down from the eye n

corrodes the skin of and near the lids. At this stage the ocular conjunctiva is also considerably swollen and hyperæmic and œdematous, and forms a shining elevated ring around the periphery of the cornea (*chemosis*). With the profuse secretion the necrotic superficial epithelium is cast off. In consequence of this the superficial bloodvessels are easily ruptured, and considerable hæmorrhage may ensue when the lids are everted.

As the formation of pus becomes more profuse, the subjective symptoms grow less severe. This is due particularly to the decrease of the swelling of the lids and conjunctiva which now takes place. If left to itself the disease will run its course in from four to six weeks. The swelling and discharge will gradually decrease and may finally disappear altogether, or a chronic blennorrhœa may remain behind.

In severe cases patches of infiltration are formed in the palpebral conjunctiva, which, as they contain no blood, are yellowish white and hard and appear like diphtheritic patches. In others the exudation forms membranes on the surface of conjunctiva as they are seen in croupous conjunctivitis.

The impaired nutrition of the cornea, as well as the direct infection from the pus, in which it is continually bathed, frequently cause corneal affections. These are usually infiltrations and ulcers, which are especially destructive when occurring during a purulent conjunctivitis. The ulcers begin for the most part near the corneo-scleral margin, and may travel around the whole periphery of the cornea (*ring-shaped ulcer*). They may, however, appear in any part of the cornea. If the disease is not brought promptly under control, the ulceration leads to perforation of the cornea, prolapse of the iris, and eventually to the total destruction of the eyeball.

Acute purulent conjunctivitis is frequently seen in the eyes of the newly-born (*blennorrhœa neonatorum*). The etiology and course of the disease is the same as in the adult. In both the infants and the adults the attacks vary in intensity, but it can be stated that, comparatively speaking, the infant's eyes can stand the ravages of this disease better than those of the adult.

In both the disease is caused by the specific virus. It lies

in the hands of the general practitioner who delivers a woman
to prevent the possibility of the development of the blennorrhœa, and on him solely must fall the blame, if he fails to do
so. We know that not only virulent gonorrhœal discharges of
the vagina, but also the mucoid discharge of a chronic vagi-
nitis is to be dreaded. It is, therefore, best to make it a
routine practice in all cases of childbirth to apply such pre-
ventative treatment as gives security against blennorrhœa neonatorum. That this can be done most effectively may be
gleaned from the statistics of larger lying-in institutions.
While formerly in some of them blennorrhœa neonatorum happened in from fifteen to thirty per cent. of the cases, these same
institutions have been able to reduce the percentage to considerably below one per cent. Of all the blind people about
one-third have been blinded for life by this same disease.
These figures may by intelligence and care be wiped out altogether. The simplest method of prevention is that of *Crédé*,
and consists of the instillation of one drop of a two per cent.
solution of nitrate of silver into the conjunctival sack of each
eye of the newly born. This destroys the virus whether it is
still suspended in the fluids of the conjunctival sack, or has
already attacked the superficial epithelial cells of the conjunctiva. The more certain will the result be if to this instillation
be added the careful washing and drying of all the parts of the
infant as they come out of the vagina. Particular attention in
this should be paid to the face (eyelids) and hands, as there
can hardly be any doubt but what infection in some cases is
brought about by particles of matter which remain adherent
to the hands and face, and are wiped into the eye a few days
after birth. The importance of this subject ought to be fully
recognized by every practitioner, and on his conscience and his
alone will and must rest all the blame for the dire results if he
is neglectful.

A very great help in the treatment of purulent conjunctivitis
and a safeguard against infection of the conjunctiva lies in the
frequent cleaning of the eye of all discharge by means of a
solution of bichloride of mercury (1 to 4 or 5,000). Mild
cases may even be successfully treated with this alone. Others
prefer the injection of liquid vaseline under the lids.

With regard to the treatment of the affection under consideration, the rule usually laid down is, to make continued ice applications alone, as long as the discharge is scant and serous and blood-stained. When the discharge becomes purulent and abundant, caustic treatment is to be commenced. The latter consists in the daily application of a one per cent. solution of nitrate of silver to the whole conjunctiva. Corneal affections are no contra-indication to this treatment, but they call for increased carefullness in confining the application of the caustic solution to the conjunctiva and also for the instillation of either sulphate of atropia or sulphate of eserine (one per cent.).

To this caustic treatment may be added other anti-phlogistic measures besides the ice applications. These are the application of leeches to the temple, or, what is of greater value, scarification of the conjunctiva. In some cases the pressure from the swollen lids is very great, and relief may be given by canthotomy.

Some surgeons have of late divided the upper eyelid into two halves by a vertical section through its whole thickness, in order to reduce the pressure on the eyeball, and to be able the better to remove the discharge.

In chronic blennorrhæa there is but little swelling of the lids. The conjunctiva, especially of the fornix is, however, often hyperæmic and infiltrated, and has a granular or velvety appearance. The discharge is thin fluid pus mixed with tear-fluid.

For the treatment of chronic blennorrhœa the daily application of the sulphate of copper in substance, is often preferable to the nitrate of silver solution.

§55. *Croupous or membranous conjunctivitis* is as well marked and distinct a form of inflammation as is the similar affection of the mucous membrane of the larynx. Its characteristic feature is the formation of a grayish white membrane on the surface of the conjunctiva, combined with slight swelling of the eyelids and a scant mucus or muco-purulent discharge. This croupous membrane may involve a part only, or it may cover the whole area of the palpebral conjunctiva. It is, however, never observed to form on the ocular conjunctiva. Small patches of such a membrane are often seen accompanying purulent conjunctivitis.

At first the croupous membrane adheres rather firmly to the conjunctiva, and can be removed with difficulty. When the affection has lasted a few days, however, it may be easily removed by rolling it up with a sponge or linen rag, but only to be rapidly reformed. The conjunctiva beneath it is very succulent, its papillæ are enlarged, and it has a bluish-red tint. The removal of the croupous membrane often causes a slight bleeding. As the affection progresses the papillary swelling increases.

The cornea during a croupous conjunctivitis is but rarely affected, although it may be partially or totally destroyed by ulceration. Still the cases, in which the croupous conjunctivitis leads to serious results are very rare.

When the croupous patches are small, continued ice-water applications and the instillation of a solution of bichloride of mercury (1 to 5000) are in place. Daily caustic treatment is indicated when the membrane can be easily detached. When the membrane covers the whole inner surface of one or both eyelids, and can only with difficulty be removed, it is best to confine the treatment to cold applications and careful cleansing. Only, when the membrane becomes loose and can easily be removed, is caustic treatment advisable.

The conjunctiva is, furthermore, sometimes the seat of a *diphtheritic inflammation.* This affection is less frequent in America, than it is, for instance, in Germany, yet it is met with from time to time, and in hospital practice it sometimes occurs epidemically. Its cause is probably *Læffler's* bacillus of diphtheria.

The appearance of conjunctival diphtheritis is so characteristic that the physician who has once seen a case, can hardly confound it with anything else.

In this form of conjunctivitis the eyelids are swollen, often stiff and very hard, so that they cannot be everted, or only with the greatest difficulty. The pain is extreme, and is aggravated by the slightest pressure. While the exudation in croupous conjunctivitis lies on the surface of the conjunctiva, in diphtheritic conjunctivitis it also fills the whole thickness of the tissue of the conjunctiva. The latter, therefore, is whitish in color and anæmic from the pressure of the exudation around the bloodvessels. The diphtheritic exudation often appears in isolated patches only. These patches appear depressed when compared with the surrounding swollen and congested conjunctiva. The diphtheritic membrane cannot be removed. The ocular conjunctiva is greatly swoolen and in rare cases it, too, becomes the seat of diphtheritic infiltration. There is a small quantity of watery secretion. An attack of diphtheritic conjunctivitis is usually attended with fever.

During this affection the cornea very rarely remains intact, and oftener is totally destroyed. The eyelids also may suffer extensively, or may even slough off altogether.

After the active stage of the disease is past, which happens in from eight to ten days, the exudation is slowly dissolved, leaving an ulcerated conjunctiva in a state of purulent inflammation. This ulceration may heal without causing any deformity, or it may leave a considerable amount of scar-tissue behind. Sometimes necrosis of a portion of a lid or the whole lid is observed.

Diphtheritic conjunctivitis is mostly found to be from the start, and to remain during its course a localized affection; a fact which has an important bearing upon the question of diphtheria in general. It sometimes extends from a diphtheritic process in the throat and nose, or begins in the eye and travels downwards, but these complications are rather rare.

The treatment consists in continued ice applications and frequent instillations of an antiseptic solution. When the exudation begins to dissolve, the application of caustic treatment, as in purulent conjunctivitis, is indicated. Any corneal affection must, of course, be cared for.

§56. *Trachoma, granular conjunctivitis, granulated eyelids*, is that form of inflammation of the conjunctiva in which in addition to swelling of the eyelids, œdema and swelling of the papillæ of the conjunctiva (*papillary trachoma*) (See Fig. 37), and an abnormal secretion, there is also a *formation of gran-*

ules. These latter are round, grayish, translucent, sago-like bodies, slightly elevated above the surrounding conjunctival surface, but embedded in the conjunctival tissue. They

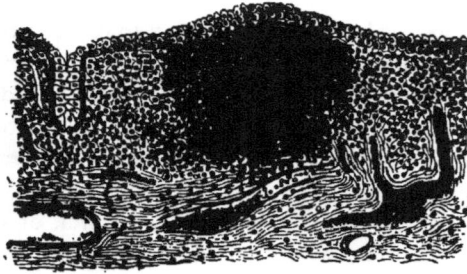

FIG. 37.—(After Saemisch). Papillary Trachoma.

are aggregations of lymphoid cells (*lymphomata*), and resemble the lymph-follicles of the intestinal tract. (See Fig. 38). Their usual seat in the beginning of the affection is the fornix of the conjunctiva, but they may spread over

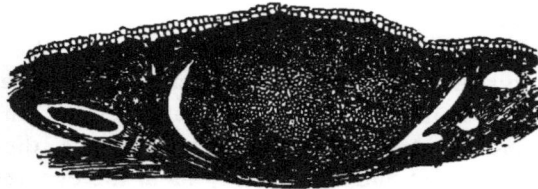

FIG. 38.—Trachoma granule (lymphoma).

the whole inner surface of the eyelids, and even to the ocular conjunctiva and corneo-scleral margin. Later on these granules undergo characteristic changes, and give rise to characteristic affections of the cornea, of the subconjunctival tissue, and of the eyelids. The presence of the granules is the characteristic feature of trachoma, although they may be partially hidden, in the beginning, by the swollen papillæ, and thus may for a time escape detection. On the other hand the characteristic results produced by this form of conjunctivitis in the conjunctiva, cornea and eyelids, enable us to make the correct diagnosis of trachoma having existed, even when the granules have entirely disappeared.

The granules are not arranged in any regular way, but are usually irregularly grouped. They vary in size. After having existed for a certain time the granules become organized, and are transformed into connective tissue, so that in the end their former seat is marked by scars in the conjunctival tissue.·

During the progress of the disease the papillæ of the conjunctiva are also swollen, and the subconjunctival tissue is often greatly infiltrated; later on this infiltration becomes organized, and the newly formed connective tissue contracts, causing shrinkage of the conjunctival sack and atrophy of the mucous glands. The tarsal tissue undergoes fatty degeneration, and by the contraction of the new-formed scar-tissue its curvature becomes gradually changed. Thus the margins of the eyelids are more and more turned inward, and the eyelashes begin to scratch the cornea (*entropium*).

By the trachoma of the conjunctiva itself, and by the ensuing entropium and trichiasis, a constant irritation of the cornea is kept up and an inflammatory reaction takes place in its tissue. This latter may progress but slowly and may lead only to the destruction of the superficial layers, or it may progress rapidly, and lead to destructive ulceration or even to sloughing of the cornea.

In the former case the superficial layers of the cornea at its upper part become dim and infiltrated, the epithelial coat loses its luster, small superficial ulcerations may appear, and gradually bloodvessels are seen to grow into the infiltrated tissue from the corneo-scleral margin between the corneal epithelium and Bowman's layer. This condition is called *pannus*, and may extend downwards over the area of the pupil, and thus render the patient virtually blind. (See Fig. 39).

In the second case we have to deal with small phlyctaenula-like ulcers springing up near the corneal periphery, or with larger ulcers, and abscesses in the corneal tissue. These latter may lead to perforation of the cornea with prolapse of the iris, and subsequently to shrinkage of the globe or to the formation of a staphyloma.

Iritis is often observed in connection with corneal affections dependent on trachoma, and must be considered as a serious complication.

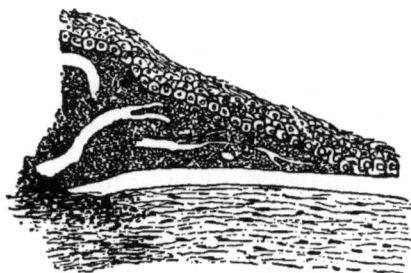

FIG. 39.—Section through pannus of the cornea, showing that the infiltration lies mainly between Bowman's layer and the epithelium.

Finally the shrinking of the conjunctival sack may attain such a degree, as to render it nearly impossible for the patient to open his eyes. At this stage the several ducts of the lachrymal glands are often obliterated and the mucous glands of the conjunctiva are atrophied. In consequence the con-junctiva and cornea are almost perfectly dry. This condition is called *xerophthalmus*.

The symptoms here described are in the main those of the *chronic form of trachoma*, which is the most frequent one.

In rarer cases we may have occasion to observe an *acute trachoma*, either as a primary affection or as an exacerbation during the progress of chronic trachoma.

Acute trachoma causes great irritation of the eye, lachry-mation and swelling of the eyelids and conjunctiva.

There may be a considerable amount of discharge, so that the affection appears to be an acute catarrhal conjunctivitis. As the swelling and discharge become less, the conjunctiva, especially that of the lower lid, is found studded with granules. Such attacks of acute trachoma may come on in eyes which have previously been in apparent health. If they come on during the progress of a chronic trachoma, the symptoms are more or less modified by the pre-existing condition. Such intercurrent acute attacks are, moreover, very apt to affect the cornea. Acute trachoma may end in recovery or it may go over into the chronic form.

In the acute, as well as in the chronic type of trachoma, the subjective symptoms are chiefly the feeling as of dust or sand in the conjunctival sack, of heat in the eyelids and inability to use the eyes for any length of time for near-work with comfort. These symptoms are particularly noticeable in the morning and evening. The discharge sometimes glues the eyelashes together, but not always.

Trachoma is rarely seen to affect one eye only. It usually affects both eyes from the beginning, or one is soon infected from the other. In chronic trachoma the subjective symptoms may for a long time be very mild, so that the patient is not even aware of his disease, until, perhaps, the eyes begin to tire when used at night, or to feel uncomfortable in the morning, or until an intercurrent acute attack brings him to the physician.

Trachoma is pre eminently a chronic disease. If left alone it may heal after many months or years. When it is healed, the conjunctiva and lids, and often the cornea, are not in their normal condition, but show changes characteristic of the disease. If no other symptoms are left characteristic bands of scar-tissue can usually be found in the conjunctiva of the upper lid.

Even under the most careful treatment it may take months and even years to cure trachoma although the most modern methods of treatment have rendered the disease much more tractable and cut short its duration to a very gratifying degree. Relapses still occur, though much less frequently, than formerly.

This most gratifying result has been brought about by modern antiseptic agents, and since the disease is directly attacked by surgical means.

Whenever we have to deal with granular formations in the conjunctiva, whether acute or chronic, or where we can suspect their presence within a swollen and hyperaemic conjunctiva, the first step in the treatment must be to render the conjunctival sack aseptic by means of repeated flushing with a solution of bichloride of mercury (1 to 3, 4 or 5,000), followed by the squeezing out and crushing of all granular (lymphomatous) formations. Since this treatment was first advocated by *Hotz*, (in his method he used the finger-nails) it has been greatly improved upon. A number of instruments have been devised among which that of *Knapp* deserves the preference, since by means of it it is possible to do this operation in a most thorough manner. However, it does not matter what instrument is being used, as long as the granules are all squeezed out or are crushed. This little operation is quite painful and it may be necessary to give a general anaesthetic in order to fully accomplish the purpose, whilst in most cases local anaesthesia by means of cocaine will help the patient to bear it. Where there is little general swelling and the granules are well pushed to the surface, this operation followed by the flushing with a solution of bichloride of mercury several times daily, but particularly night and morning, may suffice for a cure. I have seen eyes which after years of treatment had been considered fit only for enucleation, get well rapidly and become again very useful, indeed, after this operation. Ulcers and pannus disappear almost as if by magic. I cannot understand, how anyone can go back to the old methods or depend on them solely with such a simple and powerful remedy at hand. It is usually best to follow up this operation by a few weeks of treatment with the light application of sulphate of copper in substance or nitrate of silver in solution. When the corneal symptoms predominate the employment of massage with an ointment of yellow oxide of mercury, or aristol is in place. The rapidity of the improvement is apt to induce the patient to leave off treatment before a cure is established. It is best, however, to insist on it until the conjunctival swelling is perfectly removed. The patient should then keep on using the flushing with the bichloride of mercury solution for a month or so. Although relapses under this management have become rare, they do occur and I keep the patient, when it is possible, under surveillance for a prolonged period. As a source of relapses I have often recognized granules hidden in the tissue of the lachrymal caruncle and semilunar fold. Its best, therefore, when squeezing the granules out of the lids, not to forget the caruncle.

Other methods of treatment aiming at the removal of the granules have been highly recommended, but none of them is as simple and easy as the one described, although I do not doubt they are just as efficacious.

The excision of the conjunctival folds, or the destruction of the fornix by galvano-cautery are unnecessary and must be condemned since better means have been found.

If the palpebral fissure has become contracted so as to interfere with the raising of the upper eyelid, canthotomy or canthoplasty must be performed.

It seems to be a pretty well established fact that trachoma is due to the infection with a micro-organism which differs but little from *Neisser's* diplococcus of gonorrhoea, and thus, perhaps, the origin of the disease is to be traced back to vaginal secretions. Like other micro-organisms, those causing trachoma, thrive better in low, flat lands, than in higher altitudes, where the disease is consequently not so frequent. The contagiousness of trachoma is very great, and endemics are not rare. Children should, therefore, be isolated and kept from school when affected by this disease.

Trachoma is one of the most frequent of eye-diseases, and, although oftener observed among the poor, it is found in all classes of society.

Where malaria is prevalent, trachoma seems to be frequent also, and I have even heard it stated that trachoma is so intimately related to malaria that it will yield to anti-malarial treatment to the exclusion of local applications. It is almost needless to say that the latter idea is an erroneous one, although the conditions which favor malarial diseases are very much the same as those which favor the appearance of trachoma. Although anti-malarial treatment has no direct value in the treatment of trachoma, still the general debility, caused by malarial fever, may, like any other constitutional affection, render the system less able to resist disease, and for this reason only, an anti-malarial treatment may have a place in the treatment of trachoma.

When chronic trachoma had run its course and little or no granular tissue was to be found in the swollen conjunctiva, but the eye remained very irritable and pannus persisted, rendering vision poor or useless, it was with older surgeons the practice to infect such eyes with purulent discharge (gonorrhoeal pus). The results in some cases were evidently gratifying enough to allow of such a practice to be recommended. However, good these results may have been, nobody is likely, I think, to make use of such a remedy to-day. It can be the better dispensed with, since we have a remedy by which a similar inflammation may be produced in such eyes and which is free from the most objectionable features of inoculation with pus. The seeds of *Abrus precatorius* (*Jequirity*), a leguminous tropical plant, have been used for a long time in *Brazil* in the treatment of trachoma. This remedy was introduced into modern therapeutics by *de Wecker*. It may be used in the form of an infusion of the shelled and crushed seeds which is brushed on the conjunctiva two or three times until the inflammatory reaction begins. Another method, which I have practiced for years and can highly recommend, makes use of it in the shape of a nearly impalpable powder, which is applied directly to the portions of the conjunctiva where the intensest reaction is desired. The application in either form causes a croupous conjunctivitis which after a few days takes on a more purulent character, and then gradually decreasing, disappears in from one to two weeks. Its final effect is a disappearance of the swelling and roughness of the conjunctiva and the clearing up of the pannus to a very considerable degree. I have seen corneæ, which had been useless for years on account of dense pannus, clear up to such a degree, that they offered no appearance of a dimness to the naked eye, and the eyes could be used for the finest work without discomfort. Such a result seldom follows one application alone; but there is nothing to hinder from making one or more successive applications. The effect of the remedy must be carefully watched, however, during the acute stage of inflammatory reaction, as it is known to have caused the formation of deleterious ulceration of the cornea. The more vascular the pannus is, the less such a result is to be dreaded. I might almost say, the worse the condition of the cornea before the application, the better the prognosis for the results of a jequirity inflammation. After

the effects of this inflammation had perfectly passed away, I
have frequently found formerly unsuspected nests of granules
remaining behind, and have squeezed them out.

There are several milder forms of granular conjunctivitis
which are by some considered diseases of a totally different
nature on account of the different clinical picture they repre-
sent. Histologically they cannot be distinguished from the
trachoma we have described. It is, however, well to know
that the small follicle-like granules which may be found in the
lower or upper fornix, and are accompanied by annoying but
not violent symptoms (*follicular catarrh, follicular trachoma*)
are not apt to develop into the more serious form, although this
does happen. It is best for practical purposes to rid the patient of whatever granules are found in the conjunctiva in the
manner above detailed.

The treatment of the corneal affections dependent on trachoma will be spoken of in Chapter VIII.

The operative treatment of the affections of the eyelids
caused by the same disease has already been detailed in
Chapter III.

§57. Another form of inflammation to which the conjunctiva
is subject, especially in childhood, is the *phlyctænular* (so-called
strumous or lymphatic) *conjunctivitis*. It affects primarily the

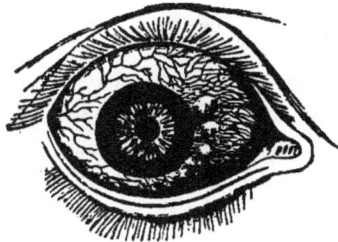

FIG. 40.—(After Dalrymple). Phlyctaenulæ of the conjunctiva.

ocular conjunctiva, and especially the limbus conjunctivæ. (See
Fig. 40). On the injected and infiltrated conjunctiva a
small papula or vesicle is formed, or sometimes several at the
same time. This vesicle contains in most cases only a serous

fluid and a few round cells, which are arranged around a terminal nerve twig. The inflammation may remain confined to the neighborhood of the vesicle or it may spread over the entire ocular conjunctiva of the eyelids. Frequently we find the same formation of vesicles also on the cornea (*phlyctænular keratitis*). By and by the vesicle bursts, its contents escape, and a small ulcer remains in its place. The ulcer may now gradually heal, or the morbid process may be continued by the successive appearance of new vesicles. In some cases this vesicle is secondarily infected by some pyogenous micro-organisms and a pustule or small abscess results (*pustular conjunctivitis*). In other cases a pannus-like infiltration takes place (*pannus scrophulosus*).

In some cases the general irritation is but slight; in many cases, however, it is very great. The eyelids are œdematous and hot; there is continued lachrymation, and such a dread of light that a child suffering from this affection will not only hide the face in the day time, whenever this is possible, but even bury it deeply in the pillow at night.

All these symptoms lead to the production of blepharitis. The skin becomes irritated and excoriated, and in warm weather, especially, the whole face may present a continuous surface in a state of eczematous inflammation. There is frequently tonic blepharospasmus combined with the photophobia.

The disease belongs essentially to childhood, and is but seldom found in the adult.

It is probably due to an infection from the mucous discharge of the nose, which is never wanting in such cases. I have made it a rule, therefore, for years to advise the parents to that effect and to have them pin a clean handkerchief daily to the child's dress, which is to be used to wipe the eyes only. The disease is more frequent among uncleanly, than among the well-kept and well-washed children.

The severer cases, and especially such as show a tendency to frequent relapses, are generally accompanied by marked signs of scrophulosis. Phlyctænular conjunctivitis is not as contagious an affection as the other forms of conjunctivitis. It leads but seldom to serious consequences, and may even get well without medical interference. Yet, as the primary cause is not easily removed, relapses are frequent, or even the rule.

Besides the general treatment, which is directed against the constitutional disorder, (iron, iodides, cod-liver oil, salt water baths, etc.,) this affection calls for vigorous local treatment. This consists in cold applications and in the daily insperga-tion of calomel or iodoform, or the use of the ointment con-taining from 1 to 4 per cent. of yellow oxide of mercury.

To this I always add the frequent flushing of the conjunctival sack with a four per cent. solution of boracic acid. Milder cases will get well with such a solution alone. The photo-phobia and blepharospasmus may be relieved by the instilla-tion of a cocaine solution. This will in most cases replace the older methods of forcibly opening the eyes or dipping the face into a basin with cold water. In fact in the treatment of phlyctænular conjunctivitis concaine plays a very important role. Not only can all treatments better and more thoroughly be applied, but the children, being enabled to keep the eyes open, are more willing to play out-doors and are more prone to stand some light, two great factors in effecting a quicker cure. Atropine need now hardly be used, excepting when deeper seated corneal affections accompany the phlyctænular conjunctivitis.

§58. During an attack of measles, or preceding it, the con-junctiva may become the seat of an inflammation (*exanthema-tic conjunctivitis*). It has usually the character of an acute catarrhal conjunctivitis and may be accompanied by consider-able discharge. It needs no special treatment, although it may be well to flush the conjunctival sack with a solution of boracic acid.

Pemphigus sometimes is observed to occur in the conjunc-tiva. It may affect the conjunctiva alone or exist at the same time in other parts of the body. Small spots are found in the conjunctiva which instead of the epithelium are covered with a grayish exudation. As these spots are gradually changed into cicatricial tissue, others appear, and in this man-ner the whole of the conjunctival sack may be destroyed and the eyelids become glued to the eyeball. The cornea thus be ing continually covered, the eyes become useless. It usually attacks both eyes. No treatment seems to be of any avail.

§59. Among the *wounds and injuries* of the conjunctiva none are of great importance, or require special treatment except burns.

Burns with gun powder, if they concern the conjunctiva only, are usually of little importance. Yet if a great many grains of powder are embedded in the conjunctiva, it is best to remove them by cutting them out, by lifting up a minute fold of conjunctiva with fine forceps and snipping it off with scissors, or by touching them with galvano-cautery.

Burns with hot water, steam or carbolic acid, though painful at first, are but rarely of much significance Cocaine instillations to allay the pain and cool compresses are usual all, that is required, and regeneration takes place in a few days.

Burns by acids or alkalies, especially by lime or by melted metals and glass, may give rise to the most disagreeable affections, through the destruction of the tissues. Lime infiltrates the tissues to a considerable depth, and thus sticks fast to them. If an eye burnt with lime is seen immediately, a careful washing out of the conjunctival sack with acidulated water (vinegar will do) may in some measure limit the destructive action, but unfortunately we seldom see such cases early enough to do much in this way.

In all cases of burns of the conjunctiva the first thing to be done is to cleanse the conjunctival sack carefully of all foreign substances which can be easily removed. This done, atropine and cocaine should at once be instilled and ice applications be made.

If the destruction extends to the subepithelial tissue of the ocular and palpebral conjunctiva, and perhaps to the cornea, the ulcerated surfaces, lying continually in close contact, may grow together, thus forming a *symblepharon.* (This may also, happen from diphteritic ulceration). In mild cases we my sometimes succeed in preventing its formation by keeping the eyelid everted as much as possible, but as a rule symblepharon will occur in spite of all our efforts. Instillations of oil into the conjunctival sack, are usually resorted to, but are of little value. I have seen better results in several cases where the melted metal could not at once be removed from the lower conjunctival sack, and where by its presence it subsequently successfully opposed the formation of a symblepharon. I am therefore inclined to think, that where the burn of the conjunctiva is caused by melted metal, which is unirritating in its nature, and is usually flattened out smoothly, its presence in the conjunctival sack might be allowed for some time under careful watching. In the same manner shields of celluloid or hard rubber have been recommended to be worn in the conjunctival sack until healing has taken place. We must, of course, be on our guard against possible injury to the cornea, which, however, is not very likely to result from the presence of a smooth, indifferent foreign body lying in the lower cul-de-sac.

§60. In rare cases the conjunctiva has been found to be the seat of a localized primary or secondary *tuberculosis*. The tubercles appear as small trachoma-like nodules which may increase in size and number, coalesce, and finally become superficially ulcerated. The presence of tubercle bacilli alone will render the diagnosis certain. Early destruction of the nodules by galvano-cautery, or their excision may bring about a cure. *Lupus* which also attacks the conjunctiva usually grows into it from the neighboring skin.

Syphilitic ulcers may also be found in the conjunctiva. They are mostly initial chancers, produced by kissing. Sometimes they appear to be tertiary (gummatous) lesions.

An *amyloid* degeneration of the conjunctiva has been seen by a number of observers. It is characterized by a yellowish, translucent swelling of this membrane, especially of its fornix. This swelling may grow until it interferes with the movements of the lids, when its removal is indicated. No other treatment seems to be of use. The chemical reaction will ensure the diagnosis.

Œdema of the conjunctiva when not due to an inflammatory condition of the lids or eyeball may be one of the symptoms of trichinosis. It is present to a slight degree in the eyes of heavy drinkers, and it may occur in patients who for some other affection are taking one or the other of the iodide salts.

Subconjunctival ecchymosis is frequently observed as a result of contusions; also after a fit of violent coughing, as in whooping cough, etc. It is harmless, and calls for no treatment; in fact, treatment is wholly unavailing to hasten the absorption of the extravasated blood, which will disappear of itself in from two to four weeks, according to circumstances. Gentle massage may, perhaps, hasten this absorption.

§61. When the palpebral and ocular conjunctiva near the corneo-scleral margin, or the palpebral conjunctiva and the cornea are grown together in the shape of a bridge, the condition is called *symblepharon anterius*. When the union has taken place farther back in the conjunctival sack and reaches to the very fornix of the conjunctiva, it is called a *symblepharon posterius*. It is clear that any such attachment between eyelid and eyeball must impede their movements. When the whole, or at least the largest part of the conjunctival sack is thus obliterated by the union of the palpebral with the ocular conjunctiva, and, perhaps, with the cornea, we speak of *anchyloblepharon*. In this condition the movements of the eyeball and eyelids are, of course, almost totally abolished; the eye also is generally so far damaged, as to be worthless as an organ of vision.

In symblepharon anterius, in which the fornix is not involved and in which the adhesion forms a bridge-like band, connecting the eyeball with the eye-lid, the simple division of this bridge is generally sufficient; but in cases of more extensive symblepharon division of the band is unavailing, unless some means can be devised to fill the gap resulting from the destruction of the ocular or palpebral conjunctiva. This may be affected by covering the ocular wound-surface by conjunctival flaps from the same eye, by transplantation of flaps from other (even rabbits) eyes, or by covering the defect on the inner surface of the eyelid by a cutaneous flap tilted over the lid-margin or even drawn through a cut through the lid (*boutonnière*) and fastened to the inside.

In cases of symblepharon of the lower eyelid it has also been recommended to keep the eyelid permanently everted after the dissection, until the wounds are healed by means of a needle run through a fold of skin.

§62. The conjunctiva, especially its ocular portion, is sometimes the seat of tumors, which may be either benign or malignant.

Lymphangiectasia, distention of lymph-channels, is sometimes seen in the ocular conjunctiva as a small transparent and shining conglomeration of bead-like bodies which lie in the subconjunctival tissue and can be moved about with it. They are seldom large enough to cause any annoyance. When they do, they may be made to disappear by puncture, or by cutting off their anterior wall with scissors.

Pinguecula, a small yellowish elevation near the corneo-scleral margin on the medial or lateral side of the cornea and in the line of the palpebral fissure, is perfectly harmless. Its name would imply that it is of a fatty nature, which, however, is not the case. It is simply condensed and histologically changed subconjunctival tissue and its formation is probably due to the movements of the eyelids. If it becomes inflamed and swollen, and thus gives rise to annoyance, besides being in some measure disfiguring, it may be removed by a clip of the scissors.

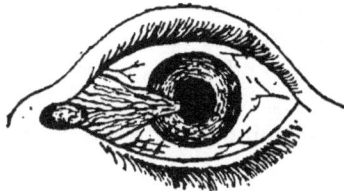

FIG. 41.—Pterygium internum of the left eye.

Pterygium (wing-skin) is a triangular fold of conjunctival tissue, widest near one angle of the palpebral fissure or the fornix, and more or less pointed at its insertion on the corneo-scleral margin or on the cornea. It is oftenest found on the nasal side, more rarely on the temporal side of the eyeball, occasionally in the direction of one or the other rectus muscle. It may for a long time remain stationary; when inflamed, however, it is apt to grow farther towards the centre of the cornea, and thus it may in time interfere with vision and even destroy it. Its presence is frequently a cause of chronic conjunctivitis. (See Fig. 41).

The formation of pterygium may be due to a marginal ulcer of the cornea, to which an overlapping fold of the nearest part of the ocular conjunctiva has become adherent. We find, therefore, in transverse sections a layer of conjunctival epithelium, incarcerated between it and the cornea or sclerotic, undergoing retrogressive metamorphosis. I have seen two cases in which this colloid metamorphosis of the incarcerated epithelial cells had caused the formation of a cyst under the pterygium, which on being punctured discharged a small quantity of viscid, colloid matter.

As it is, however, often seen to develop without any visible ulceration, there must be other causes which lead to it. A mechanical explanation, if I may call it so, was offered by *Young*, which I think is an excellent one, for the formation of internal pterygium as seen so frequently in farmers, seamen, firemen, engineers and railroad employees, in short among men exposed to rough weather, dust and heat. When trying to shut the eye to keep off these irritants, the orbicularis muscle is contracted tightest at the temporal side, while near the nasal side enough of it is left comparatively relaxed in order to have a small triangular opening to see through. It is then, of course, on that part of the conjunctiva that all the contents of the air will be deposited, and that is just where pterygium occurs most frequently.

Theobald tries to account for internal pterygium by the hyperæmia produced by the use of the internal recti during near work. Unfortunately, pterygium is but rarely found among the class of people who use their eyes for near work continually, while it is frequent among those classes that rarely, if ever, read.

Fuchs takes the position that every pterygium originates in a pinguecula, an opinion which *Horner* and others held before him. That a pinguecula may develop into a pterygium cannot be doubted, but a pinguecula alone does not make a pterygium. There must be other forces at work, for whose applied energy the pinguecula may form an especially favorable point of attack. According to *Poncet's* theory these are micro-organisms.

Whatever the ultimate cause may be, the pterygium grows into the corneal tissue in the shape of a wedge, which raises the superficial layers, *Bowman's membrane* and the epithelium. In this growth the conjunctiva is dragged along in such a manner as to impede the movements of the eye in the opposite direction.   In internal pterygium the semilunar fold disappears and the lachrymal caruncle may be dragged a considerable distance from its original situation and towards the cornea.

Pterygium shoul be removed as soon as it begins to encroach on the cornea or causes continued irritation.

A number of methods have been devised for accomplishing this.   In spite of a thorough operation relapses or newly formed pterygia are occasionally met with.

In the improved method of *Knapp's* the pterygium is first carefully dissected off the cornea and sclerotic and then by a horizontal cut divided into an upper and a lower half, one of which is to be stitched into the lower and the other into the upper fornix.

The gaps into which these halves of the pterygium are to be stitched are formed by making an incision through the conjunctiva from the base of the pterygium upward and downward into the fornix.

Another method is that of *Galezowsky*.   After the pterygium has been thoroughly dissected off the cornea, the tissue towards its base is undermined with the scissors.   Finally a thread armed with two needles is carried through the apex of the pterygium so as to form a loop, and its ends are brought out through the conjunctival tissue at the base of the pterygium.   When the threads are tied, the apex is folded under so that the pterygium is doubled on itself.   This causes at first the appearance of a disfiguring swelling, but as the pterygium atrophies this swelling disappears.

Prince advocates simple evulsion of the pterygium with the forceps.   He claims for this somewhat barbaric method, that the cornea is clearer when all is healed than by any other method.

The oldest method is to excise the tissue of the pterygium in the shape of a rhomboid, after it has been well dissected off the cornea and sclerotic.   A partial closure of the resulting gap in the conjunctiva by undermining this membrane upwards and downwards and stitching the wound-lips together, may follow the excision. I have for years performed this operation, preceded by a thorough flushing of the conjunctival sack with a bi-chloride of mercury solution, and followed by a cauterization of the corneal wound with pure carbolic acid. The patient, when discharged, is further directed to keep on using the solution of bi-chloride of mercury for several weeks longer. Although I have operated on a large number of patients for pterygia of all sizes, relapses after this method have been extremely rare.

The conjunctiva near the corneo-scleral margin is sometimes the seat of a congenital growth of *dermoid* tissue. The little tumor usuall encroaches upon the cornea. It consists of all the elements of the skin, including hair. When annoying by its appearance or on account of irritation caused by the hair, it should be, and is easily, removed.

Congenital sub-conjunctival lipoma is sometimes found with dermoid or separately.

Cystic formations, not of a lymphangiectatic nature, are sometimes met with in the conjunctiva They may be easily excised, when they cause any annoyance.

Granulomata, (*polypus, proud flesh,*) sometimes spring from the conjunctiva during inflammatory conditions and after injuries or operations. They may be pedunculated or have a broad base. Their removal with the scissors is simple and may or may not be followed by cauterization of the wound.

The malignant tumors found in the conjunctiva are either epitheliomatous or sarcomatous in character. When either of these originate in the conjunctival tissue, they begin at or near the limbus of the ocular conjunctiva. In its growth the epithelioma gradually advances on the cornea spreading first between *Bowman's* layer and the corneal epithelium. Later on *Bowman's* layer is perforated and the growth spreads into the corneal tissue proper. It may lead to perforation of the cornea and thus find a way into the iris and into the deeper portions of the eyeball. It may also reach the interior of the eyeball along one or more of the larger-blood vessels which pierce the sclerotic.

The sarcomata are mostly pigmented. They grow much in the same manner as that of an epithelioma, and sometimes, while spreading into the corneal tissue, the sarcoma is also seen to grow around the corneo-scleral margin. (See Fig. 42).

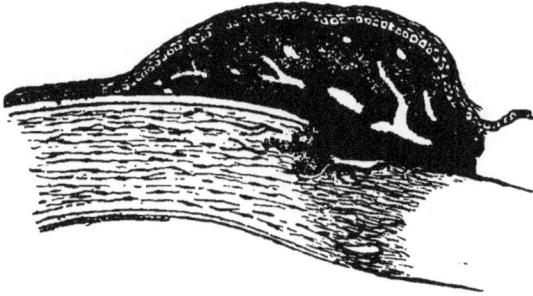

FIG. 42.—Pigmented episcleral sarcoma encroaching upon the cornea. The growth has spread quite a distance towards the center of the cornea under the epithelium and without destroying Bowman's layer. It is now entering the corneal tissue proper at the corneo-scleral margin.

It is possible in the early stages to remove these tumors successfully, when they have not yet pierced Bowman's layer. Their destruction by galvano-cautery may be recommended when the cornea is not yet deeply implicated. Later on the eye has to be sacrificed.

CHAPTER VIII.—DISEASES OF THE CORNEA.

§63. All forms of inflammation of the cornea, *keratitis*, cause a dimness of a part of, or even of the whole area of, the cornea. This dimness is due to the infiltration of the corneal tissue with cells which wander into it, or are newly-formed within it. Such an infiltration may either become absorbed or lead to the formation of pus. It may lie superficially or concern the deeper lamellæ alone, or it may pervade the whole thickness of the cornea. The resulting dimness may disappear altogether later on or remain permanent.

Keratitis is usually accompanied by a symptomatic conjunctivitis which may be but barely perceptible or be quite a prominent symptom. Iritis may be, and often is, seen as a complication. During the absorption blood-vessels are often seen to grow into the corneal tissue. They spring from the terminal loops in the periphery of the cornea or from the scleral blood-vessels (see Chapter I). Such newly formed blood-vessels may, during the healing process or even afterwards, atrophy and disappear; in a minority of the cases they remain persistent. Keratitis may attack one eye only, some forms of it nearly always affect both eyes, although not always at the same time.

As a general rule, the danger of a complicating iritis makes it necessary in cases of keratitis to at once instill atropine and to keep the pupil dilated. In a few exceptional cases eserine, which causes the pupil to contract, may for the time be indicated, especially when the intra-ocular tension is increased and a perforation of the cornea is to be feared. To allay pain, which is sometimes very severe in corneal affections, the bathing of the eye with hot water, the use of steam thrown against the lids by an atomizer, or of the Japanese hot box, and the instillation of a 4 per cent. solution of cocaine are to be relied upon. Cocaine, however, should be used moderately. Its too free use has a bad effect on the nutrition of the cornea, and consequently on the process of repair.

§64. *Phlyctaenular keratitis (keratitis phlyctaenulosa, lymphatica, strumosa)*, is essentially the same affection as phlyctaenular conjunctivitis, and as has been stated under that head, the one is very often seen associated with the other. (See Chapter VIII). The treatment is exactly the same, and need not be again insisted on. The affection often leaves no trace behind, only where the phlyctaenular ulcer has involved Bowman's and some deeper layers, a slight scar is formed, which later on appears as a small gray spot (*macula*). In other cases in which a leash of blood-vessels extends from the periphery towards the seat of the phlyctaenula (*fascicular keratitis*), these may remain for some time after the heeling of the phlyctaenula. The blood-vessels usually disappear later on, but often leave a dimness of the cornea in their place.

As in phlyctaenular conjunctivitis the phlyctaenula may become secondarily infected with pyogenous germs. The result is a larger pustule (*keratitis pustulosa*), which gives the disease a graver aspect.

Sometimes a number of phylctaenulæ, appearing like miliary tubercles, are formed at the same time or in quick succession near the corneo-scleral margin, but this form of miliary phlyctaenulæ differs from the other form in no other way.

Although the treatment of phlyctaenular keratitis is the same as that of phlyctaenular conjunctivitis, we have to add, that in the keratitis it is best, as a rule, to instill atropine. The tonic treatment of the general system should never be forgotten in these cases (syrupus ferri iodati, cod-liver oil, salt water baths, etc.) The disease frequently recurs, but the tendency to its recurrence dies out near the age of puberty. In rare cases it is observed in the adult.

§65. There are several other forms of keratitis characterized by the formation of small vesicles which must not be confounded with phlyctaenular keratitis. In febrile diseases and sometimes during an attack of *herpes zoster ophthalmicus*, small vesicles are seen to spring up on the surface of the cornea. The eye is then in a state of considerable irritation and sometimes very painful. As the vesicles burst small ulcers result and by the confluence of such ulcers the ulcerated surface may appear branched in several directions (*keratitis dendritica*).

Keratitis bullosa is another form of inflammation of the cornea in which vesicles are formed. This disease, according to *Landesberg*, is primarily an inflammation of the deeper portions of the cornea with the formation of an exudation below Bowman's layer, or between it and the epithelium. The affection is accompanied by great irritation and pain and an increased intraocular pressure. These symptoms subside when the vesicle bursts or is opened. Relapses and intercurrent exacerbations are the rule. Treatment seems to be of little use.

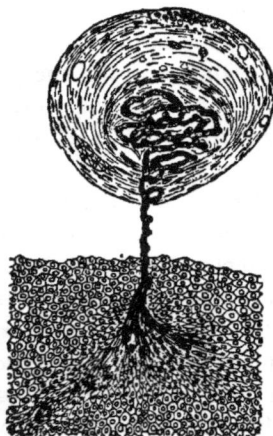

FIG. 43.—(After Nuel.) Keratitis filamentosa. Microscopic appearance of a thread-like epithelial body, and the region of the cornea from which it springs.

Of late a number of observers have described a form of superficial keratitis in which small thread-like formations are seen to hang from the cornea (*keratitis filamentosa*). Microscopical examinations have made it clear that these threads which were thought to be fibrine, are in reality excrescences of the corneal epithelium which grow with a peculiar twist. (*Hess, Nuel*). (See Fig. 43).

§66. Keratitis affecting particularly the deeper seated portions of the cornea is called *parenchymatous* keratitis (*keratitis
interstitialis, punctata, profunda, syphilitica, scrofulosa*). This
affection may be ushered in by a state of irritation, photophobia and slight pain. In other cases all such premonitary symptoms are wanting. The next symptom is the appearance of a
grayish spot lying usually in the middle or deeper lamellæ of
of the cornea. This may appear at the corneo-scleral margin
and gradually spread over the whole cornea, or it may start in
the center of the cornea and gradually spread towards its
periphery. The spreading may take place by the appearance
of new gray spots which later on more or less coalesce, or the
original gray spot may simply grow in size. However, even
when it seems to be one solid gray spot, it can be seen by the
aid of a magnifying glass, that in reality it consists of numerous smaller ones. The spreading of the dimness over the cornea usually takes from several weeks to several months, and depends somewhat on the severity of the attack. During this
period the epithelium suffers, too. Its natural lustre is destroyed, and it looks steamy or stippled and irregular.

When the dimness appears first at the corneo-scleral margin
it may be soon entered by a bunch of small blood-vessels
coming from the terminal loops of the corneal periphery.
These vessels follow the dimness for some extent into the corneal area. Aside from these, other blood-vessels (sometimes
only one, apparently) spring from the scleral vessels and grow
into the infiltrated portion. In some cases they are so numerous that at the height of the process the cornea appears reddish like raw flesh. While in some cases the deeper tissues of
the eye are not affected, in others hyperaemia of the iris,
iritis and irido-cyclitis may occur. When the whole of the
cornea is dim, sight may be almost abolished. When the dimness has started in the center it often leaves a clear space at
the corneal periphery, which, however, is useless for vision.

While the symptoms of photophobia, pain and lachrymation
are quite prominent in some cases, they are mild and totally
absent in others.

After the disease has been at its height for some time, the
cornea in some place near its periphery begins to clear up.

Other peripheral portions follow suit and finally the center also gets clearer and clearer, until but barely a thin cloudiness or not even this is visible to the naked eye. During and after this clearing process the newly formed blood-vessels also disappear, at least to the naked eye. With a magnifying lens, even in the apparently perfectly clear cornea, some dimness and fine branches of blood-vessels may be recognized many years after an attack of parenchymatous keratitis (*Hirschberg*). Sight, in spite of this, is nearly normal.

In rare cases the infiltration may lead to pus formation (abscess or ulcer) and even to perforation of the cornea, and I have seen an anterior polar cataract to result from such an occurrence. In other rare cases the cornea gives way to the intraocular pressure and is bulged out.

Sometimes the cornea does not clear up as well as above described. In these cases the infiltration has led to the formation of scars (connective tissue) and sclerosis of the cornea. When such a sclerosis has taken place in the whole area of the cornea, this membrane grows flatter and anterior phthisis of the eyeball may follow. According to the situation of such a scar or scars, vision may be partially or very greatly impaired. In such cases an artificial pupil may still be the means of giving the patient useful vision.

Parenchymatous keratitis is always a chronic and tedious affection. There are variations in the severity of the attack, but on an average it takes from three to six months to run its course; some cases take much longer. The duration of the affection may, however, be shortened by intelligent treatment, especially if the case is seen at its beginning.

The disease may attack one eye alone, but, as a rule, the other eye is also invaded by it. However, both eyes are not frequently attacked at the same time; there may be even an interval of one or several years between the affection of the first and that of the second eye. It is more frequently seen in children than in adults.

Through the influence of *Hutchinson's* researches it has become the rule, particularly with the English authors, to call this form of keratitis simply syphilitic keratitis, hereditary syphilis being looked upon as its sole cause. The fact is unquestioned that a large number of the patients suffering from this disease show evidences of inherited syphilis, as Hutchinson's teeth, scars at the angles of the mouth, enlarged lymphatic glands, partial loss of hearing and others, yet the affection is also observed in quite a number of cases in which no such characteristic symptoms are found. Some of these patients may be of a strumous or an otherwise enfeebled and anæmic constitution, others appear to be perfectly healthy. Occasionally parenchymatous keratitis is met with in cases of acquired syphilis.

The prognosis with regard to the clearing up of the corneal dimness is rather good, and is the better the greater the vascularization of the dim cornea. The local treatment, therefore, must include chiefly such remedies as are likely to bring about and stimulate the new formation of blood-vessels in the corneal tissue. This is most successfully accomplished by the frequent application of moist heat. Hot bathing of the closed eye three or four times during the day, for from half an hour to an hour at the time should at once be ordered. In its place steam from an atomizer may be used. If there is photophobia, smoked glasses are to be worn. To prevent or render less nocuous any iritic complications, atropine must be instilled. In order to hasten the recuperative process mild massage with an ointment of yellow oxide of mercury (two to four per cent.) once a day is highly to be recommended. The cornea may be slightly touched with sulphate of copper, iodoform or calomel may be dusted into the eye, although in my experience these remedies have not given any better satisfaction. In the later stages and in obstinate cases I have seen good effects from the instillation of a two per-cent. solution of creoline (which, however is very painful) and from the application of the galvanic current. Subconjunctival injections of a few drops of a one per mille solution of bichloride of mercury in cases in which syphilis is undoubted seem rational and are highly recommended.

To this local treatment constitutional treatment must be added, where it is called for, and it may be well in all cases to give some tonic or iodide of potassium. It must, however be kept in mind, that whenever an iodide is exhibited internally it is secreted with the tears, and for that reason no calomel must be applied to the eye, as it would cause the formation of the iodide of mercury, which produces undue irritation or even caustic effects.

§67. When an infiltration of the cornea and a local necrosis lead to the formation of pus within the corneal tissue we call it an *abscess of the cornea.*

FIG. 44.—Abscess of the cornea with hyopyon (hypopyon-keratitis).

This affection appears always in an acute form. We see in the cornea a dim yellowish spot, which may be near the surface or lie embedded in the deeper layers. It is usually round or semilunar, or it may be ring-shaped. Its outlines are, however, never very sharply defined, as the surrounding tissue is also in a state of infiltration. If the affection progresses the nearest surrounding parts become also necrosed. When such an abscess has reached a certain size, pus cells will wander from it into the anterior chamber and there fall to the lowest point, and form what is called *hypopyon.* (See Fig. 44). In the formation of this hypopyon the iris and ciliary body, which frequently become also inflamed, may take a part. The aqueous humor becomes generally turbid. The abscess may,

furthermore, increase so as to break through the anterior surface, and thus form an ulcer. Sometimes it breaks through
into the anterior chamber. The pus may also become absorbed with or without the newformation of blood-vessels.
The cavity of the abscess is then filled with newformed connective tissue and a gray spot will be left to mark its former
location. In very rare cases the pus 'is absorbed without any
newformation of connective tissue, or at least with not enough
to fill the cavity. On the other hand the abscess may cause perforation of the whole thickness of the cornea and lead to anterior synechia, loss of the crystalline lens and vitreous body,
staphyloma, and total loss of the eye.

Abscess of the cornea is a very painful affection, the pain
being apparently greatest at night and keeping the patient
from sleeping. There is great irritation, photophobia and
lachrymation. The disease appears, as a rule, in one eye only.
It is seen more frequently in old and debilitated individuals
than in young persons. It is due to a pyogenous infection of
the cornea and frequently occurs after slight injuries to this
membrane (with subsequent pyogenous infection), especially
when there is a pre-existing affection of the tear-passages.

The prognosis in this affection is always doubtful, as we
cannot predict how far the process will extend. As a rule,
however, its progress stops when the abscess has perforated
the surface. If there is a great quantity of pus in the anterior
chamber, the prognosis is, of course, more doubtfel than if
there is little or none.

The treatment consists in hot moist applications to the
closed eye, as hot as the patient can bear them. This and the
instillations of cocaine will help to relieve the otherwise often
excruciating pain. To prevent iritis atropine should be instilled. Instead of the instillations of atropine, eserine has of
late been used frequently, but seems to have not only no advantage over the atropine but it even seems to favor the formation of iritis.

If, under this treatment, the pain and the process of the
affection do not stop, and voluntary rupture or absorption do
not readily occur, I have seen great and immediate benefit,
not from opening the anterior chamber, as many do, but from

cutting simply through the layers anteriorly to it into the
cavity of the abscess and allowing the pus to escape, just as
one would open an abscess elsewhere. *Saemisch's* method,
which is frequently practised, consists in cutting through the
whole thickness of the cornea about in the middle line of the
abscess, and letting the aqueous humor and pus escape through
this. The incision should begin and end in healthy tissue, and
must be reopened from day to day until the formation of
pus ceases. This method is very painful, does not give
quick results, and in the end the case often does no better
than it would have done with less interference; anterior synechia, moreover, follows very frequently. The simple opening
of the abscess is more rational and fully as effective.

The most rational method is to destroy the anterior wall of
the abscess and the seat of the abscess itself by galvano-
cautery. This must be followed by instillations of a solution
of bichloride of mercury (or the inspergation of iodoform, or
the instillation of a solution of pyoctanine) and closure of the
eye. Thus we may succeed in rendering and keeping the
affected territory aseptic, and to put it in the best conditions
for a rapid healing. If carefully done the cauterization arrests
the spreading of the abscess and will leave no greater scar,
than the abscess would have left.

If there is a lachrymal affection present it must, of course,
be attended to.

§68. *Ulcers of the cornea* are either caused by a previous
abscess in the way just described, or the infiltration is at first
superficial, leading presently to necrosis of the epithelium and
most superficial layers of the cornea.

They are due to the presence of a micro-organism, mostly
of a pyogenous character, without a preceding traumatism, or
following one. Corneal ulcers are among the most frequent
affections of the eye and vary greatly as to situation, form
and progress. They may come on when there is no other eye-
affection present, or they may be directly due to another
affection of the eye, particularly to diseases of the conjunctiva, trichiasis and entropium. The ulcer may make its

appearance at the center or at the periphery of the cornea, or at any intermediate portion. (See Fig. 45).

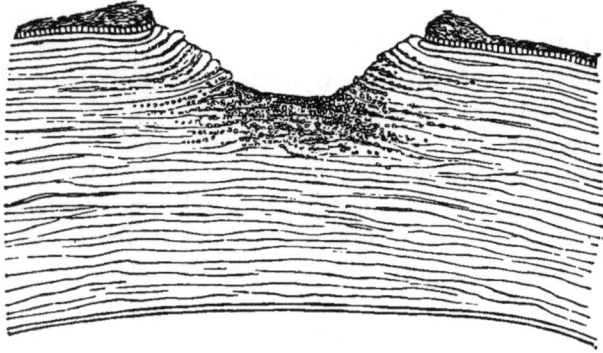

FIG. 45.—Histological appearance of an ulcer of the cornea. Its ground and walls are infiltrated with round cells. The epithelium surrounding its margin is in a state of proliferation.

As long as the ulcer is progressing its walls and fundus, as well as the surrounding parts, are grayish or yellow from infiltration. The pus cells may also invade the anterior chamber and form a hypopyon. Iritis is also a frequent complication. When the ulcer heals, its walls and fundus first become clear, and then, with or without the formation of blood vessels in the cornea, the process of repair begins. The ulcer is gradually filled up with new-formed translucent connective tissue, which becomes covered by epithelium, thus leaving behind it a gray spot. Often the progress of the ulcer does not stop until it has gradually eaten through the whole thickness of the cornea and caused a perforation. Before this occurs, Descemet's membrane, which is very resistent, may sometimes be seen to bulge forward through the ulcerated part (*keratocele*) for some time. When the perforation takes place the aqueous humor escapes, the iris prolapses into the corneal wound, and the ulcer may heal with an anterior synechia. If the perforation takes place in the center of the cornea the anterior capsule of the crystalline lens comes in contact with the ulcer and plugs it. This may give rise to the formation of an anterior polar cataract. In other cases these accidents may lead to the loss of the crystalline lens, and even of the vitreous body, to the formation of a partial or total corneal staphyloma, or to loss of the eye through shrinkage. In rare cases the loss of substance by ulceration is not repaired and its site is covered with epithelium only (*fascette*).

Like abscesses of the cornea, ulcers are usually very painful, and cause photophobia and lachrymation.

Their occurence is not particularly confined to any period of life. They frequently appear, as has already been stated, during conjunctival affections. Corneal phlyctaenula and ulcers occur also in connection with malarial fever, and a particular form of ulcer has been described as *malarial keratitis* (*Kipp* and others).

The prognosis is doubtful and depends on the size and locality of the ulcer.

The most rational, quick and efficient way of treating a corneal ulcer is by galvano-cautery (or actual cautery). Cocaine renders this little operation almost painless. Care must be taken to cauterize the whole ulcerated surface, even should a perforation occur. This is to be followed by the instillation of a solution of bichloride of mercury or pyoktanine, or by the inspergation of iodoform, or similar antiseptic remedies, and closure of the eye by a bandage or plaster. The cauterization destroys the injurious germs and is at the same time an excellent stimulus to the newformation of tissue which is to replace the lost tissue. Pure carbolic acid may take the place of the galvano-cautery.

Very small superficial ulcers may be treated by means of hot bathing or steaming of the closed eye, flushing with a solution of bichloride of mercury or of boracic acid, or inspergations of iodoform, and gentle massage with an ointment of aristol or yellow oxide of mercury. It may be advantageous to keep the eye closed with a bandage or plaster in the intervals between the local application, and it may be necessary to keep the pupil dilated by the instillation of atropine.

It is the duty of every practitioner to avoid and prevent the use of any eye-wash or bathing lotion containing acetate of lead, a practice which is still too common in this country. When such lotions are applied to the ulcerated cornea, in-

crustations of lead take place on this membrane, which interfere greatly with vision and can frequently not be removed
later on.

Whenever it is for some reason impracticable to cut short
the process of ulceration by means of the galvano-cautery and
the ulcer refuses to cleanse itself, and continues to invade new
territory and causes the formation of hypopyon, the operation
introduced by *Saemisch* is applicable and is here of greater
utility than in cases of abscess of the cornea. Care must be
taken to begin and end the cut in the healthy tissue, as far as
this may yet be possible.

In other cases simple paracentesis of the cornea will be sufficient, or an iridectomy may be called for, as the intra-ocular
pressure is sometimes increased (*secondary glaucoma*).

Ulcers complicating a conjunctival inflammation seldom
need special treatment. (See Chapter VII).

In paralysis of the trigeminus, ulcerations (sometimes abscesses) of the cornea are often observed. These cases are
easily recognized from the fact that the sensibility of the cornea is greatly reduced or totally abolished. Usually the secretion of tears and of mucus is also diminished and thus a dryness of the corneal epithelium results. The reduced sensibility
makes it possible for small foreign bodies to wound the cornea
or even remain on it, without apparently causing discomfort.
The reduced secretion of tears in itself gives rise to superficial
excoriations.

Similar ulcers occur in facial palsy, *Basedow's* (*Graves'*) disease, during the semi-unconscious state of febrile diseases,
when reflex winking is abolished, or when from whatever cause
the upper eyelid cannot protect the cornea from air, heat,
dust and micro-organisms.

If it is impossible to cure the paralysis of the trigeminus,
or to remove the disability of the upper lid to protect the
cornea by any other means, this must be done by shortening
the palpebral fissure by the operation of tarsorraphy. (See
Chapter III.).

§69. As has already been stated, any process in the cornea
which is attended with destruction of tissue and which neces-

sitates repair by means of newly formed connective tissue, must result in a scar. Scars, in contrast with the normal, transparent corneal tissue, are only translucent, and therefore appear as more or less dense grayish, or even white spots which according to their situation, may or may not interfere with vision; when large and centrally placed they may render the eye partially or even practically blind by their density, or they may give rise to irregular astigmatism, by altering the curvature of the cornea. (See Fig. 46).

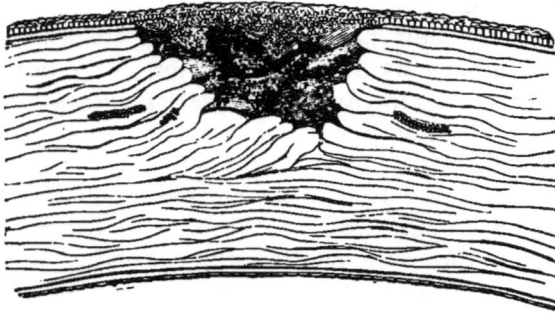

FIG. 46.—Histological appearance of a healed ulcer of the cornea (leucoma). The former loss of substance is filled with new formed, irregularly arranged and but semi-transparent connective tissue. Bowman's layer being absent, the epithelium has grown downward into the scar-tissue.

These scars may clear up to a certain extent, especially when situated superficially, and particularly in children's eyes. This process of clearing may sometimes be very beneficially influenced by treatment. Gentle massage with an ointment containing yellow oxide of mercury, or iodide of potassium, or blue ointment, the instillation of tincture of opium diluted with water, and spraying the cornea with steam or solutions of sulphate of copper or of tannic acid, have been recommended and are in in some cases beneficial.

In certain cases patients suffer from very annoying dazzling from light, which is irregularly refracted in passing through such a translucent scar, especially if it covers only a part of the pupillary area. In such cases it is sometimes advisable to tatoo the scar with India ink so as to render it impermeable to light. Tatooing may also be used for a simply cosmetic effect, when a scar of the cornea is very disfiguring.

When the scar lies in front of the whole pupillary area, and renders the patient virtually blind, an iridectomy may often restore useful vision. This should be made preferably in a place where the upper eyelid is not likely to cover it, but the direction of the iridectomy is usually determined by the position of the clearest and best part of the remaining cornea, and thus may have to be made where the surgeon would least desire, were he left free in his choice. An iridectomy for this purpose should be made as small as possible, as the patient will see better through a small pupil, which does not allow many irregularly refracted rays to enter the eyeball, than through a larger one which admits them.

The grayish zone seen in the periphery of the cornea, usually in people of an advanced age, is called the *arcus senilis*. It is not scar-tissue, but the result of a fatty degeneration of the corneal tissue and cells, and is of no importance.

FIG. 47.—Staphyloma of the cornea.   On the right hand iris and cornea are firmly attached to each other, atrophied and stretched.

As has been before stated, an abscess and ulcer of the cornea, which leads to perforation, may result in the formation of a *staphyloma*. The staphyloma consists in a partial or total bulging of the remains of the cornea together with the iris, which in these cases always adheres to it. If left alone, the bulging process may go on until the eyelids can no longer be closed over the eyeball. Such eyes are seldom free from irritation, and they are liable to be attacked by various forms of inflammation; sometimes a secondary glaucoma results. (See Fig. 47).

In the beginning of a partial staphyloma the instillation of a solution of eserine or pilocarpine, and frequent puncturing of the bulging tissue may be successfully resorted to. The latter procedure causes more and more scar tissue to be formed, which by its shrinking often brings about a flattening of the protruding parts. Sometimes an iridectomy or sclerotomy proves successful in these cases, combined with the application of a compressive bandage. I some cases when the staphyloma is very small it may be best to cut or burn a part of it away with the galvano-cauter.

To remove a total staphyloma of the cornea an abscision of the whole cornea may be made, combined with the removal of the crystalline lens, if it is still present. The margin of the opening thus left in the sclerotic will generally heal together, and a good stump on which an artificial eye can be worn is the result.

To hasten the closure of the wound *Critchett* stitched the scleral wound-lips together, the needles passing through the ciliary body. *Knapp* improved on this procedure by suturing the conjunctiva only, the threads acting like the strings of a tobacco-pouch.

In modern times evisceration of the eyeball, emptying it of all its contents with a sharp spoon and subsequent insertion of a glass ball (*Mules*) have been recommended in order to improve the stump left after abscision of the staphyloma.

Still, as any such stump may become inflamed, and cause sympathetic inflammation of the good eye, it is better, as a rule, to remove every eye with total staphyloma of the cornea, whenever the case cannot be kept under continued observation.

When the whole of the corneal tissue is changed to scar tissue and no perforation has occurred, it may still be possible to restore some sight, by the implantation of a transparent flap of corneal tissue (*Hippel*). In order to do this a disc of opaque corneal tissue is removed by means of a trephine down to Descemet's membrane, which latter must remain intact. Into this gap a disc of transparent cornea is transplanted, and may remain transparent enough to allow of some useful vision.

Conical cornea, a cone-shaped bulging of one or both corneæ, may be the result of the loss of resisting power with or without previous ulceration and gradual thinning out of the corneal tissue. (See Fig. 48).

FIG. 48.—Keratoconus.

The abnormal refraction which by the gradual increase of the cone gets worse and worse, disturbs vision very considerably. Hyperbolic glasses may sometimes be worn with success by such patients (*Ræhlmann*). *Dor* recommends contact lenses. The production of a retracting scar by cauterizing the cone at its apex, or the excision of the apex, has sometimes had a beneficial effect on the sight of eyes so affected.

§70. *Injuries to the cornea* are of frequent occurrence. Very frequentiy they are complicated by injuries to the deeper parts of the eyeball. When the cornea alone is injured the conditions are comparatively simple. In all such cases it is necessary to establish as soon as possible a condition of relative asepsis by all means at our hand, even the galvano·cautery, and to maintain it.

Injuries may be inflicted by blunt or by cutting instruments, by heat, or by chemicals.

Abrasions of the corneal epithelium are often seen. They are painful and cause lachrymation, especially when the eyeball is moved. Flushing with an antiseptic solution, instillations of atropine and cold or warm applications with rest, which may be brought about by a compressive bandage, will allay the disagreeable symptoms, and the defect will generally be healed in from one to two days.

Aseptic cuts that do not penetrate the whole thickness of the cornea heal very readily.

When the corneal tissue is perforated by the injuring body, just as in the case of a perforating ulcer or abscess, the aqueous humor will escape, and, as the posterior parts are thus moved forward by the intra-ocular pressure, the iris may be caught between the wound-lips, or it may even prolapse through them, and be held in that position. As prolapse of the iris increases the danger to the eye, and may even be the means of exciting inflammation of the uveal tract or the formation of staphyloma, the iris should, if possible, be set free. If it is impossible to cause its retraction by slightly rubbing the cornea with the eyelids, by the instillation of the sulphate of atropia or of eserine, or by instrumental help, the last resort is to cut off the protruding part and thus permit the remainder of the iris to retract within the eye. Some surgeons suture the corneal wounds. *Burns* of the cornea by hot foreign bodies or by chemicals, especially lime, cause a more or less superficial necrosis of the corneal tissue with subsequent ulceration. When the cornea alone is injured, the treatment will be the same as in the case of a cut of the cornea, except when the burn resulted from lime. In this case every particle of lime should, if possible be removed and the remainder neutralized by acidulated water (vinegar). It sometimes happens that we find a shell of lime lying on the cornea after all inflammatory symptoms have passed off. It may be possible to remove this, and the patient's sight may thus be greatly improved.

When the burning material has at the same time injured the conjunctiva there is danger of the formation of a symblepharon, which must, if possible, be prevented. This point has already received the necessary attention. (See Chapter VII).

The tumors of the cornea take their origin almost invariably from the adjacent conjunctival tissue, and therefore only secondarily invade the corneal tissue. They have been spoken of in Chapter VII.

CHAPTER IX.—DISEASES OF THE SCLEROTIC.

The sclerotic proper is not very apt to become inflamed, and such inflammatory symptoms as sometimes occur in it take their origin probably in the episcleral tissue.

§71. *Episcleritis*, or *scleritis*, usually affects but one eye at the time and is a localized inflammation. Near the limbus of the cornea and beneath the highly hyperæmic conjunctiva, a purple elevation is seen, which is often painful and tender on pressure. The deeper the seat of the inflammation the deeper is the purple color. Although beginning as a small localized tumor, the swelling often wanders around the whole periphery of the cornea. Episcleritis may run its course without further complication, or it may be complicated with affections of the cornea, iris, choroid and even of the retina. It is a tedious disease, often resists treatment for a long period, and is very apt to recur.

The reason for these peculiarities lies in the fact that its occurrence is mostly due to some general diathesis, rheumatism, gout, or syphilis being present in the majority of cases. Another class of cases occur in females at the climacteric period, or when suffering from some trouble of the sexual organs. Episcleritis is also occasionally due to an injury, in which case it yields more readily to treatment than when it is of constitutional origin. The direct cause of the non-traumatic cases of episcleritis is, as *Mooren* states, probably a pathological condition of the blood-vessels brought about by some diathesis.

The treatment which appears to be most successful consists in hot bathing, instillations of atropine and the use of an ointment of aristol or yellow oxide of mercury combined with massage. In cases of syphilitic (gummatous) episcleritis inspergations of calomel and sub-conjunctival injections of a solution of bichloride of mercury seem to act better than any other remedy.

Whenever a general diathesis is present its treatment must accompany the local applications.

In a few cases that have come under my observation at a very early stage, the use of the muriate of pilocarpine, either hypodermically or instilled into the eye, has been followed by a remarkably rapid recovery. It has been recommended also to cut trough the swollen part down to the healthy sclerotic, or even to scrape the whole swelling off with a sharp curette.

§72. There is an insidious chronic form of scleritis affecting the deeper layers of the sclerotic and complicated with inflammation of the uveal tract, which is for the most part noticed by its results only, namely, the formation of a *scleral staphyloma*. Such a stayhyloma forms a bluish elevation which usually begins at one of the weaker parts of the sclerotic, especially where it is pierced by blood-vessels, and may gradually grow to a considerable size. The seats of predilection of scleral staphyloma are in the equatorial, or the ciliary region of the eyeball. In some cases the whole sclerotic may become staphylomatous (*total staphyloma*).

FIG. 49.—Ciliary staphyloma of the sclerotic. The atrophied ciliary body adheres to the sclerotic and the two are together bulged outward.

At the seat of the staphyloma the sclerotic and uveal tract are firmly adherent to each other, and become together more and more attenuated and stretched. Although the disease may at first interfere comparatively little with vision, it leads gradually to further alterations in the tissues of the eyeball, and frequently gives rise to secondary glaucoma, or perhaps, ultimately to sympathetic inflammation of the fellow-eye. Scleral staphyloma is sometimes caused by an injury, but in most cases the etiology is obscure.

According to its seat we distinguish between anterior scleral staphyloma (*ciliary, intercalary staphyloma*), equatorial staphyloma and posterior scleral staphyloma. (See Fig. 49).

The last form, the *posterior scleral staphyloma*, is due to an attenuation and stretching of the sclerotic adjoining the optic nerve entrance and generally in the direction of the macula lutea. As in this form of staphyloma there is always elongation of the antero-posterior axis of the eyeball, myopia, (short-sightedness) is always present. (See Fig. 50).

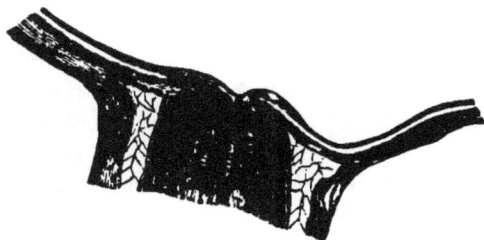

FIG. 50.—Posterior staphyloma of the sclerotic from a myopic eye. Choroid and sclerotic adhere to each other, are atrophied and bulged outward.

Total staphyloma of the sclera (*hydrophthalmus, buphthalmus*) is due to the thinning and stretching of the whole sclera, and it is usually combined with enlargement of the cornea (*megalocornea*). This condition may be congenital or acquired and is due to an abnormally high intraocular pressure and a weak scleral structure. The media may be perfectly clear. In a number of cases the cornea is dim, or the iris and crystalline lens are adherent to each other and pressed forward towards the cornea (probably as the result of a former ulceration and perforation of the cornea) and there may be signs of deeper inflammatory processes. Vision accordingly may be comparatively useful or be totally abolished. The disease may be arrested by iridectomy, or removal of the whole iris (*Noyes*); in some cases during infancy the progress stops spontaneously. The disease may attack both eyes, but occurs often in one only.

In ciliary and aequatorial staphyloma when they are not too far developed, partial abscision or an iridectomy may sometimes be useful. In most cases, however, the time for such an operation has elapsed before the patient seeks help. If such an eye is irritable, painful and unsightly and, as is mostly the case, is useless as an organ of vision, it is best to enucleate it.

§73. *Wounds and ruptures* of the sclerotic when aseptic, may heal by first intention. It is, however, well to render the eye and conjunctival sack as aseptic as possible by antiseptic remedies, and then to either stitch the conjunctival and scleral wound-lips or the conjunctival wound-lips alone together. Under antiseptic closure the healing progresses well and may leave no untoward symptoms behind. I have even seen extensive complicated wounds of the sclerotic with prolapse of the choroid, retina and vitreous body, heal well and apparently give no further trouble. Yet, as the wound-lips and deeper parts may have become infected when the injury occurred, a matter which we have no means immediately to recognize, such complicated wounds must be looked upon as something very dangerous, both for the wounded eyeball itself and its fellow. This is more especially the case when the ciliary region is involved. The physician, should, therefore, be extremely guarded in giving a prognosis in such cases.

If the wound is due to a small foreign body which is retained in the vitreous body it should be removed if possible. Particles of iron or steel may sometimes be successfully removed by means of a magnet.

When a deeper infection and pus formation has once taken place within the eye, which is usually due to the presence of some septic foreign body within the vitreous body, it may perhaps still be possible to remove the foreign body and bring the acute inflammation to a standstill, yet the future welfare of such eyes is always doubtful.

If we do not succeed in finding the foreign body and in removing it, or if we cannot stop the progress of the septic inflammation, the eyeball should be removed. This is the more imperative as such an eye is a continued menace to the fellow-eye by inducing sympathetic (*migratory*) ophthalmia. (See Chapter XVIII).

Tumors of the sclerotic are, as a rule, of conjunctival origin; the sclerotic proper is seldom, if ever, the primary seat of a new formation.

§74. Inflammation of the iris is a rather frequent affection, and is easily recognized in its severer forms or later stages. It is chiefly in its beginning that it is frequently confounded with catarrhal conjunctivitis. It may be well, therefore, to make it a rule to dilate the pupil by instillation of a one per cent. solution of sulphate of atropia in all doubtful cases.

The different forms of iritis are chiefly recognized by their products, and we have four typical forms, namely: plastic, serous, purulent, gummatous (and tubercular) iritis.

Every form of iritis is characterized by hyperæmia of the episcleral blood-vessels, and secondarily of the superficial vessels of the ocular conjunctiva.

The iris loses its luster and is changed more or less in color (a blue iris appearing greenish), a change which is most easily recognized when only one eye is affected. by comparing it with the fellow eye. Excepting in certain cases of serous iritis, the pupil is small and immoveable.

Except in slight cases every attempt to open the eye in the light is attended with profuse lachrymation, and generally with sharp pain. Iritis is, as a general rule, a very painful affection, and the pain is usually constant. It may be confined to the eyeball itself, or it may irradiate into the supra-orbital and infra-orbital regions; it is usually severer at night, and is often extremely distressing. Besides this spontaneous pain we generally find that the eyeball is tender on pressure, which shows that the ciliary body is also implicated. The eyelids are often slightly œdematous, though never to a very high degree, except in purulent iritis.

Iritis may be due to a trauma, but is more frequently a spontaneous disease.

Spontaneous iritis generally appears in one eye at a time, but very often the fellow eye is attacked while the disease is still active within the first eye.

If iritis is not properly treated in time, it always leads to attachments between the iris and the anterior lens-capsule (*posterior synechiæ*). These are easily detected by the instillation of atropia, under whose action the pupil will assume a very irregular shape, according to the number and extent of the synechiæ. (See Fig. 51). When there is a circular attachment between the pupillary margin of the iris and the anterior lens-capsule, the pupil will remain absolutely unchanged (*circular synechia*).

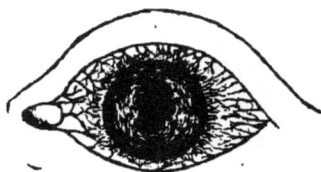

FIG. 51.—Posterior synechiæ in plastic iritis. The pupil is partially dilated by atropine.

Vision is always impaired in iritis, although to a varying degree. This is dependent on the size of the pupil, and the quantity and quality of the inflammatory products in the anterior chamber, and on the anterior lens-capsule. Moreover, as in iritis the ciliary body and choroid, also, are hardly ever free from inflammatory activity, exudations from these parts into the vitreous body help to impair vision. Such exudations are frequently observed long after the other symptoms have entirely disappeared. Some forms of iritis are very apt to recur, especially when it has been impossible to break all the synechiæ.

§75. In *plastic iritis* (*fibrinous, rheumatic iritis*) which is the most frequent form, the exudation from the iris is of a fibrinous nature. This is deposited first at the pupillary edge of the iris, then on its posterior surface, and sometimes also on the anterior surface. It glues the iris to the anterior capsule of the crystalline lens, and thus renders it partially or totally immoveable. If the disease goes on, a perfect membrane may be formed in the pupillary space, into which blood-vessels may grow from the adjacent iris (*occlusion of the pupil*). By the formation of such a pupillary membrane or of a circular synechia, or by both together, the anterior chamber may be shut off from the parts of the eyeball lying behind the iris, and thus the slow current of the fluids within the eyeball, which goes from behind forwards, is seriously obstructed (*seclusion of the pupil*). (See Fig. 52). This causes an increase of

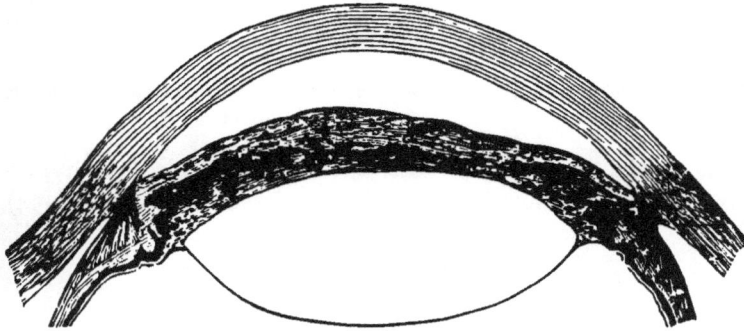

FIG. 52.—Newly formed connected tissue, the result of a fibrino-plastic iritis, unites the posterior surface of the iris with the anterior lens capsule.

tension in the posterior parts of the eyeball and the periphery of the iris, if not also glued down to the lens, may be pushed forward so as almost to touch the cornea (crater-shaped pupil). That such a condition cannot exist long without seriously endangering the function of the eye, is obvious. Most frequently the inflammation spreads backwards upon the ciliary body and choroid and leads to perfect destruction of the eyeball by a chronic iridochoroditis. In other cases the eye is destroyed by secondary glaucoma. In these stages of the disease the formation of the cataract is a frequent complication.

Luckily the inflammation may sometimes spontaneously come to an end before the more serious consequences are developed.

From the foregoing it may be seen that is of the greatest importance to recognize the disease early.

It may, therefore, be well to enumerate again the symptoms of iritis, as contrasted with those of conjunctivitis. In conjunctivitis the hyperæmia of the blood-vessels is confined to the mucous membrane of the eyeball and eyelids. (See Fig. 53). In iritis, although there may, too, be a considerable hyper-

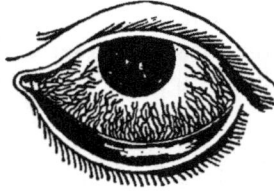

FIG. 53.—(After Dalrymple). Hyperæmia of the conjunctival bloodvessels in conjunctivitis.

æmia of the conjunctival vessels, there is, in addition, hyperæmia of the episcleral (ciliary) bloodvessels (See Fig. 54),

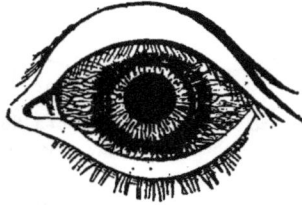

FIG. 54.—(After Dalrymple). Hyperæmia of the conjunctival blood-vessels and of the episcleral and ciliary blood-vessels around the periphery of the cornea in iritis.

which shows as a bluish-red zone around the corneo-scleral margin, over which the hyperæmic conjunctiva can be moved. In conjunctivitis the appearance of the iris is unchanged, and the pupil dilates promptly when the eye is shaded, except when the iris is hyperæmic; in iritis the appearance of the iris is materially altered, and it is inactive. In conjunctivitis, if there is pain, it is usually located in the eyelids; in iritis there is nearly always pain, and it usually irradiates into the regions surrounding the orbit. In conjunctivitis, vision is only momentarily impaired (by mucus lying on the cornea, which can be wiped off with the eyelid); in iritis sight is considerably, often very greatly, impaired. Finally, in conjunctivitis there is mucoid or muco-purulent discharge; in iritis there is no discharge, except of tears.

If an iritis has been mistaken for a conjunctivitis, the symptoms will remain unchanged, or, more probably, become worse, so long as the patient is treated for conjunctivitis; if the wrong treatment is persisted in, permanent injury to the eye, or even loss of vision may be the result.

Plastic iritis is mostly an acute disease, and may in some cases get well without proper treatment; but very seldom, without leaving its traces behind in the shape of a posterior synechia.

In some cases of plastic iritis due to trauma or syphilis, which are characterized by hæmorrhages into the tissue proper of the iris and even the anterior chamber and between the lamellæ of the cornea, a special form of exudation is found in the anterior chamber, which is termed *spongy* exudation (*Gunn, Gruening*).

This exudation formed probably by the blood-plasma looks like dim gelatine, and is often only recognized after having been partially dissolved and absorbed. It may then appear as a grayish, lens-like body, lying in the anterior chamber like a dislocated crystalline lens, but leaving its uppermost portion free. Gradually the whole of this exudation is dissolved and absorbed. This may be accomplished within a week, but it sometimes takes two or three weeks.

The average duration of a plastic iritis, even under treatment, varies from three to six weeks. Cases that come under treatment at the very beginning may, however, recover much more rapidly.

§75. *Serous iritis* is that form of iritis in which the exudation is considered to be chiefly of a serous nature. Besides the general symptoms of iritis, which are, for the most part, mild and not clearly defined, we find usually a slight increase of tension due to the spreading of the disease to the posterior parts of the uveal tract. The synechiæ formed in serous iritis are usually not very firm, nor is the pupil as apt to be small, as in plastic iritis. When the tension is increased the pupil is usually wide. During the progress of the affection, the posterior layers of the cornea become involved, and the endothelial cells of the membrane of *Descemet* proliferate. These

proliferations form numerous small whitish or pigmented dots
on the posterior surface of the cornea. The dots appear
mostly in the lower part of the cornea, and form a dimly defined triangle with its base downwards, and its point upwards
in the region of the pupil. From this peculiar arrangement it
was formerly thought that the dots were simply deposits of
fibrine from the exudation in the aqueous humor, and it is
not impossible that small fibrinous deposits may in reality
be the primary cause for the proliferation of the endothelial
cells.

If the disease progresses, the synechiæ become firmer, and
the result may be again a circular synechia of the pupillary
edge combined with bulging of the periphery of the iris. (See
Fig. 55).

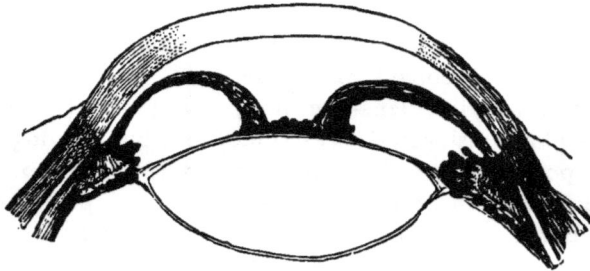

FIG. 55.—Crater-shaped iris (exclusion and occlusion of the pupil). The pupillary
edge of the iris and the iritic pupillary membrane adhere to the anterior lens
capsule. The unchecked exudation of aqueous humor into the posterior
chamber has pressed the periphery of the iris forward toward the cornea.

This form of iritis is rather chronic in its course, and does
not yield redily to treatment, which should be, in the main, the
same as in plastic iritis. When the disease has run its course
and the media are again clear we not seldom find atrophic
patches in the choroid.

§76. The characteristic feature of *purulent iritis* is that pus
cells fill the tissue of the iris and are exuded into the anterior
chamber. The iris in this form of inflammation often has a
yellowish tint which may appear in localized spots. The pus
in the anterior chamber may be fluid, or semi-fluid when fibrine

is mixed with it. It sinks to the lowest part of the anterior chamber, changing its place when the head is tilted to one side, and forms what we have already seen in ulceration or abscess of the cornea, a hypopyon. If the affection is not a part of a general purulent inflammation of the eyeball (pan-ophthalmitis), recovery may take place, but not without leaving its traces behind. The pain accompanying this form of iritis is usually very severe.

Anatomically very similar, though clinically distinct, is the fourth form of iritis the *gummatous iritis.* In it the infiltration with round cells is more localized, and we see small tumors

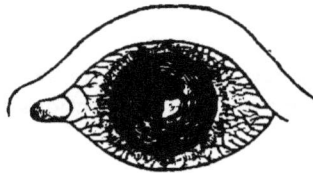

Fig. 56.—Gumma of the left iris. The pupil is partially dilated by atropine.

(*gummata*) forming mostly near the pupillary edge. (See Fig. 56). Generally we find only one, sometimes several such tu-mors. They gradually increase in size and become yellow. At this stage they may disappear again. Later on we find in their place atrophied iris-tissue and posterior synechiæ.

In other cases the gumma seems to burst through the iris into the anterior chamber and may totally fill it. This exuda-tion must, however, not be confounded with the spongy exu-dation described above. Gradually the gummatous growth is absorbed and an atrophic portion of the iris and a posterior synechia mark its original seat.

Gummatous iritis generally occurs in but one eye, more rare-ly in both at the same time.

The bacillus of tuberculosis (*Koch*) is sometimes the cause of a *tubercular iritis.* The diagnosis is difficult and remains un-certain until the characteristic micro-organism is demonstrated.

The etiology of the plastic and serous forms of iritis is not always clear. However, in a large percentage of the cases syphilis is the primary cause. Some authors state this to be the case in fully sixty per cent., and if inherited taints could always be traced, the proportion would probably be still much larger.   When iritis, as it often does, is seen in from four to six weeks after the primary sore of acquired syphilis has made its appearance, and while the characteristic skin eruptions and throat affections are present, the diagnosis is, of course, easy. In the same manner a gummy tumor developing during iritis will set our doubts at rest.   Very frequently, however, iritis is one of the later or even latest affections in syphilis, and may appear when no other syphilitic symptom can be detected. Rheumatism and gout predispose to iritis.   In quite a number of cases, probably a larger one than is known, iritis is seen to follow gonorrhœa, or to appear soon after the local infection has made itself known.   It is usually preceded by gonorrhœal rheumatism of the knees and other joints.   Iritis, also, sometimes develops during an acute infectious disease like typhus and pneumonia.

Purulent iritis is almost always due to an injury with subsequent purulent infection, and is one of the unfortunate sequelæ of unclean operations on the eye.   It may also follow an incarceration of the iris after an extraction of cataract.

In the treatment of iritis the main point must always be to prevent the formation of posterior synechiæ or to rupture those that have already been formed.   This is accomplished by the forcible dilatation of the pupil by means of mydriatic drugs, among which the sulphate of atropia holds the first place.   A one per cent. solution of this drug is to be instilled into the eye every half hour, or even every quarter of an hour, if necessary, until the pupil is well dilated (ad maximum, if possible). Those who are not often called upon to use atropia, are, as a rule, afraid to use it strong enough and often enough, and frequently when the pupil is nicely dilated they get frightened, lest it should stay so or even burst when going on with the instillations.   This is all wrong, and the fact is, that only the persistent use of a strong solution (one per cent.) will accomplish anything.   The pupil will resume its normal coudition (if there are no synechiæ) in from one to two weeks after the last instillation has been made.

Most people will bear such instillations for a prolonged period without any disagreeable symptoms, except, perhaps, a dry throat. A small number of patients, however, will be found to be most sensitive to the use of atropia, and to show signs of poisoning after a comparatively small number of instillations. The face becomes flushed, the pulse becomes rapid, feeble and fluttering, the patient is nauseated, has hallucinations and sometimes even becomes delirious. In such a case morphine must be given, the use of atropia be discontinued, and extract of belladonna may be tried in its stead. Before, however, changing the remedy, we should try to prevent the instilled solution from running down the tear-duct, by turning the patient's head in the opposite direction and closing the canaliculi for a few minutes after the instillation by pressing a finger against them.

Some eyes grow more painful when atropine is instilled, which is probably due to an increase of the intra-ocular pressure.

Sometimes also atropia proves irritating to the conjunctiva and causes a follicular swelling, which may become very disagreeable. In such cases it may again be well to change the remedy. I am inclined to think that this form of conjunctivitis is due to some infection rather than the action of the drug itself. I certainly have never seen it, since I have taken the precaution to dissolve the atropia in a four per cent. solution of boracic acid.

The next point to be considered is the relief of the often almost maddening pain, which does not always yield even when we have succeeded in fully dilating the pupil. An old and undoubtedly good remedy for this in many cases is the application of from 3 to 6 leeches to the temple. I have also often seen immediate relief following the use of cold compresses or even of a small ice-bag; in some cases bathing with hot water has been more successful in abating the pain. These remedies may be combined with the internal exhibition of antipyrine, antifebrine or morphine.

Where there is the least idea of a specific origin in a case of iritis, it will be best to at once give mercury in some form or other. Inunctions rapidly pushed, calomel in small and frequent doses, and protoiodide of mercury, or corrosive sublim-

ate may be used. If the syphilitic infection has taken place
many years before the appearance of the iritis, the use of
iodide of potassium is very effective. Even in cases where a
syphilitic origin cannot be traced, mercury has often a very
beneficial influence.

To insure a more rapid and local effect repeated subconjunctival injections of a few drops of a one per mille solution
of bichloride of mercury are highly to be recommended. Subcutaneous injections of muriate of pilocarpine, or the exhibition of decoctum Zittmanii are sometimes useful, and some
authors have reported excellent success from the internal use
of salicylate of soda or salol, especially in rheumatic iritis.
My experiences in this direction have been very disappointing.

The photophobia, which may become increased when the
pupil is dilated, calls for the wearing of smoked glasses. Most
patients feel better when staying in a darkened room and even
in bed. The bowels should be kept open.

§77. *Injuries* to the iris are in most cases complicated by
injuries of the cornea and sclerotic, and often of the crystalline lens or even of the deeper parts of the eye. The conditions and the treatment necessarily vary very much in different cases and will, as far as not here referred to, be spoken of
in Chapter XV.

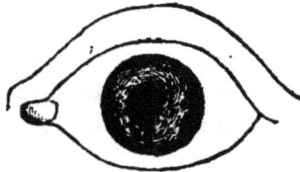

FIG. 57—Iridodialysis. The iris is torn from its ciliary insertion on the right and,
thus a second pupil is formed.

A contusion of the eyeball may result in an injury to the iris
without further complication. It may cause one or more ruptures
of the circular fibres at the pupillary edge of the iris, and the
formation of small colobomata. In other cases it produces a
detachment of the iris from its peripheral insertion (*iridodialysis*). (See Fig. 57). The former injury is comparatively rare

and is usually of little importance. In the latter there is at first considerable hæmorrhage into the anterior chamber, and after the blood is absorbed we find at the seat of the rent, at the periphery of the iris, a new, abnormal pupil, which varies in size according to the extent of the detachment. Through this second peripheral pupil rays of light enter the eyeball, as well as through the normal pupil, giving rise occasionally to some confusion of vision. When the rent is very large, so that the loosened iris floats about with the movements of the eyeball, it may be closed by forcing the iris to grow to the corneo-scleral tissue. This can be done by drawing it, by means of forceps, into and allowing it to be held between the lips of a small corneo-scleral section.

§78. The iris is sometimes the seat of *newformations*. They are either cysts or sarcomata.

The *cysts* are the result of an injury and may be of a serous nature, in which case they may be caused by the adhesion of a fold of iris-tissue to the posterior surface of the cornea. In most cases, however, these cysts originate from epithelial cell grafts, which have been forcibly driven into the anterior chamber and have grown on the iris-tissue. Both kinds of cysts may attain to a considerable size before they come under our observation. The only remedy is their total removal from the eye by iridectomy.

Sarcoma of the iris may be pigmented or unpigmented. The diagnosis is rather difficult. Yet, when we find a small solid tumor in the iris, which is steadily growing, and perhaps causes pain and increase of tension, the diagnosis of a sarcomatous newformation will probably be correct. In a very early stage such a tumor may, possibly, be removed by iridectomy with reasonable hope of saving the patient's eye as well as his life. Later on the eye must be sacrificed to save life.

§79. As functional disorders of the iris we have to mention *mydriasis*, a condition in which the sphincter pupillæ has lost its contractility and the pupil remains in a state of maximum dilatation; and *miosis*, in which the pupil is strongly contracted.

Mydriasis (when not caused by a drug) is, as a rule, only a symptom of further disorders, and especially of disorders in the nervous apparatus, and it may be due to ptomaine poisoning.

Miosis, if not caused by the action of a drug, is almost always a pathognomonic symptom of affections of the spinal cord. In these cases the contracted pupil will still further contract during the act of accommodation and convergence, but not upon the stimulus of light (*Argyll-Robertson* pupil).

Hippus, an alternate contraction and dilatation, is a rare symptom of some cerebral diseases.

CHAPTER XI.—DISEASES OF THE CILIARY BODY.

The close anatomical relation which exists between the three parts of the uveal tract makes it impossible for an inflammation to exist for some time in the ciliary body without involving the iris or, later on, the choroid. The forms of inflammation which are observed in the ciliary body are, therefore, also essentially the same as those which occur in the tissue of the iris.

The symptoms of cyclitis are mainly those of iritis. There is hyperæmia of the conjunctival and episcleral (ciliary) bloodvessels, impairment of sight, photophobia, lachrymation and severe pain, either spontaneous or on the slightest pressure on the ciliary region. The way in which a patient will rapidly withdraw his head upon such pressure, is almost characteristic of cyclitis.

§80. *Plastic cylitis*, the most frequent form of inflammation of the ciliary body, may be acute, but it is usually a chronic affection. It is characterized by the exudation of fibrinous or plastic material into the posterior chamber (see Chapter I) and the anterior portion of the vitreous body. After some time this fibrinous substance becomes organized, connective tissue is formed, and cells and bloodvessels grow from the ciliary body into this newly formed, cyclitic membrane. As this membrane shrinks, the crystalline lens is pushed forwards, and finally the posterior part of the ciliary body, with the adjacent choroid and retina, becomes detached from the sclerotic. During this process the iris and the crystalline lens become glued together, and the eyeball is destroyed by shrinkage. (See Fig. 58).

In *serous cyclitis* the exudation is mainly serous in character. This form of cyclitis is never recognized, without a co-existing serous iritis. There is increase of intra-ocular tension and all the symptoms of serous iritis are observed.

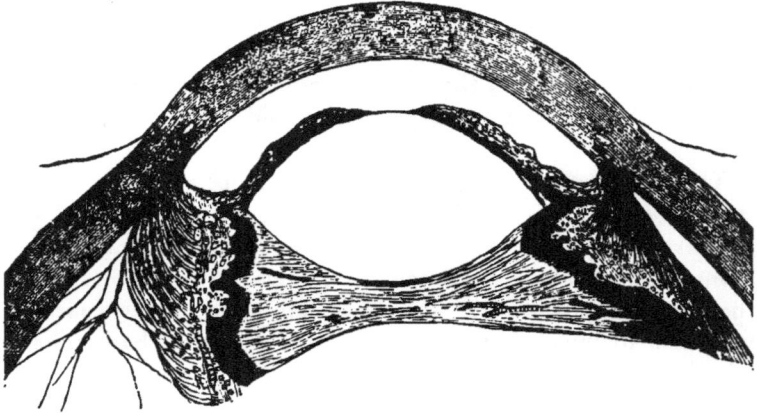

FIG. 58.—The results of fibrino-plastic cyclitis. The crystalline lens and iris are pressed forward toward the cornea by the shrinking of the cyclitic membrane running across the eye from ciliary body to cil'ary b dy. On the left, the ciliary body has become detached from the sclerotic. The posterior chamber is obliterated.

Purulent cyclitis is usually seen in cases of panophthalmitis. It is almost always due to an injury. In some cases the yellow pus exuded from the ciliary body may be seen in the anterior portion of the vitreous, when the other membranes of the eyeball have only yet begun to become inflamed.

Gummatous cyclitis is but seldom recognized, although it surely is not of very infrequent occurrence. It may remain a localized affection of the ciliary body, without doing further dammage; the gumma may break outwards through the sclerotic, or it may be absorbed. Isolated gummata have been found a number of times in the tissue of the ciliary body.

The treatment of cyclitis is in no way different from that of iritis.

Cyclitis in all its forms is but rarely a genuine primary disease; it is generally due to an injury or to syphilis. Experience has shown that an eye which has been destroyed by cyclitis due to sepsis, is a most dangerous companion to its fellow.   Sympathetic ophthalmia is very frequently the result of this affection.   It is therefore best, as a rule, to remove such an eye in time by enucleation, unless the patient is so situated that he can be kept continually under observation.

Newformations start, it seems, but rarely from the ciliary body.   They are *sarcomata* and, chiefly, *melano-sarcomata*. (See Fig. 59).

FIG 59 —Primary melanotic sarcoma of the ciliary body beginning to invade the iris.

Injuries of the ciliary body are, as has already been stated, of a very serious nature.   They will be further discussed in Chapter XVIII.

The diseases of the choroid, except purulent choroiditis, like those of other structures which make up the posterior portion of the eyeball, can only be correctly diagnosticated by the use of the ophthalmoscope. Without its aid the old by-word still holds good, that in diseases of the back-ground of the eye, the patient can see nothing, nor the physician either.

The chief symptom usually complained of in the diseases of the back-ground of the eyeball is a partial or total loss of sight, sometimes combined with photopsiæ.

The forms of inflammation which we meet with in the choroidal tissue, correspond very much with those found in the iris and ciliary body.

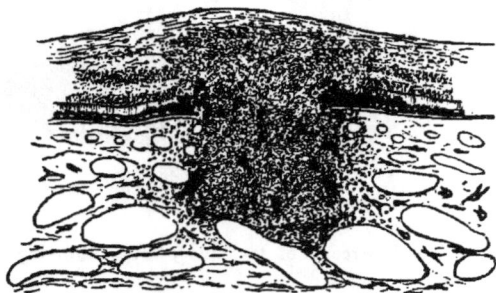

FIG. 60.—Plastic choroiditis. Histological appearance of a focus of infiltration which spreads into the adjacent retina.

§81. *Plastic choroiditis* has a number of clinical names, which, however, are all based upon varieties of one and the same pathological process; thus we speak of *exudative, atrophic, disseminate, syphilitic, peripheral, areolar choroiditis* and *central choroido-retinitis.*

In all these forms we find the exudation of fibrino-plastic material into the tissue of the choroid and the adjacent retina.

This exudation is preceded by hyperæmia of the choroid, and is combined with cloudiness of the vitreous body and congestion of the bloodvessels of the optic papilla and retina. (See Fig. 60). The cloudiness of the vitreous body may be diffuse, or separate smaller and larger flocks of fibrinous substance may be seen floating about in it.

The fibrino-plastic material exuded into the choroid and retina may be absorbed again, but in most cases it becomes organized, and when retraction of the newly formed connective tissue takes place, an atrophic spot in the retina and choroid results, which is devoid of bloodvessels and pigment and through which the whitish sclerotic can be seen. The pigment, which has been set free by the destruction of the pigmented cells of the choroidal tissue, and of the cells of the pigmentary epithelium, is collected at the periphery of the atrophic spot in such a manner as to give it a darkly pigmented, irregular outline. (See Fig. 61).

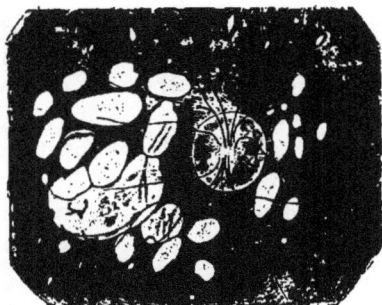

FIG. 61.—(After Foerster). Ophthalmoscopic appearance of atrophic patches in the choroid (choroiditis disseminata).

These atrophic spots are perceived by the patient as blind or dark spots (*scotomata*) when they lie near the center of the retina. While they are being formed patients are often annoyed by light-flashes, firy sparks, etc. (*photopsia*).

At least one, in most cases several, of these spots are formed during an attack of plastic choroiditis, and frequently both eyes are affected at the same time, or one soon after the other.

In the most frequent form of plastic choroiditis, the atrophic

spots lie, as a rule, near the periphery of the choroid and retina, and therefore peripheral vision is chiefly disturbed.

In central choroido-retinitis the exudation takes place in the
macula lutea (yellow spot) itself or in its immediate neighborhood. In the beginning of this form of choroiditis central vision is indistinct, and objects are seen distorted and often apparently smaller (*micropsia*) or larger (*megalopsia*) than when
seen with the healthy eye (*metamorphopsia*), straight lines appear bent or notched, etc. Finally central vision is entirely
abolished.

In both these forms of plastic choroiditis hæmorrhages into
the choroidal tissue may also occur.

Plastic choroiditis is very apt to recur, and thus to render
vision more and more defective. Yet, from the fact that it always appears in patches, which leave healthy tissue between
them, it is not apt to cause total blindness.

Syphilis, rheumatism and gout are often the basis on which
such a plastic choroiditis is developed. It happens also, comparatively often, in women at the climateric period. It is, however, seen most frequently in short-sighted eyes, and in elder-
ly people, and the short-sightedness and age must therefore
predispose to such inflammation.

With this exudative choroiditis we must, however, not confound other atrophic changes in the choroid of myopic eyes,
which are evidences of and due to the stretching of this membranes. They are pathognomonic of progressive myopia, and
will be spoken of further on in Chapter XX.

Later on the chroiditis may help in the production of cataract.

If plastic choroiditis is treated in its first beginning, especially in syphilitic cases, it may yield perfectly to treatment.
This consists in rest in a darkened room, local depletion and
the use of iodide of potassium or of some form of mercury
internally or by subconjunctival injection. Subcutaneous injections of muriate of pilocarpine are also useful. In older
cases bichloride of mercury, taken in small doses and for a
prolonged period, seems to have a very beneficial action. In
some cases in which the exudation into the vitreous body has
been the predominant symptom, and in which other treatment

has proved ineffective, I have seen very good results from the use of electricity in the form of the constant galvanic current.

The eyes must be protected from the irritating influences of light by the wearing of smoked glasses. Instillation of atropine is to be recommended, and total rest of the eyes must be insisted upon.

§82. *Serous choroiditis*, when not combined with serous iritis and cyclitis, is seldom, if ever recognized, unless it causes glaucomatous symptoms (increase of the intraocular tension). Its chief result is synchisis (liquefaction) of the vitreous body, either wholly or in part. The vitreous body, which is of a jelly-like consistency in the normal state, becomes in this affection liquid like water. Serous choroiditis may also give rise to detachment of the vitreous body from the retina, or of the retina from the choroid.

§83. *Purulent choroiditis* is characterized by the infiltration of the choroidal tissue and the vitreous body with pus-cells. It is usually a very acute affection, and thus produces symptoms which are plainly visible without the aid of the ophthalmoscope. The inflammatory process hardly ever remains confined to the uveal tract, but soon spreads over nearly all parts of the eyeball, and extends to *Tenon's* capsule (*panophthalmitis*). It then causes œdema and swelling of the eyelids and ocular conjunctiva (*chemosis*), and swelling of the orbital tissue with consequent exophthalmus. The disease, almost without exception, leads to total destruction and shrinking of the eyeball.

It, is, as a rule, very painful; in some cases, however, the pain is but slight. In the acute form the pus may break through the sclerotic or the cornea, and thus escape. This acute form may pass over into a chronic form in which the inflammatory symptoms are much less severe, yet new exacerbations will recur and recur, and it may be many years before such an eye will be apparently quiet. In such eyes we find nearly always occlusion and seclusion of the pupil, a cyclitic membrane, cataract and detachment of the retina from shrinkage of the vitreous body. (See Fig. 62). Sometimes the choroid is hypertrophied and of many times its normal thickness. The majority of its bloodvessels is obliterated. The optic nerve is atrophied and the posterior parts of the sclerotic are considerably thickened. During the progress of this affection deposits of lime and the formation of bone-tissue within the choroid may take place. Gradually the eyeball shrinks more and more (*chronic irido-choroiditis, phthisis bulbi*).

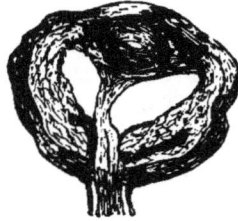

FIG. 62.—Chronic iridochoroiditis after injury. There is phthisis anterior and the whole eyeball is shrunken. The sclerotic is thicker than normal and folded. The anterior and posterior chambers are filled with new-formed tissue and pressed forward together with the cataractous lens by a shrinking cyclitic membrane, to which the detached retina adheres. The choroid is changed to a loose tissue of many times its normal thickness. The dark portions within it near the optic nerve are new-formed bone.

Purulent choroiditis is due to septic infection brought into the eye by an injury, or as the result of septicæmia (*metastatic choroiditis*). Puerperal septicæmia is especially apt to lead to it.

In most cases it owes its origin to an injury with the subsequent presence of a foreign body within the eyeball. It may lead to sympathetic ophthalmia in the fellow-eye and will, therefore, be further spoken of again in Chapter XVIII.

Treatment is usually unavailing as regards the cure of this form of choroiditis, and it should therefore be simply addressed to the relief of the more important symptoms. The pain, which is usually unceasing and very distressing, may be sometimes alleviated by cold applications, but, as a rule, the patients prefer hot ones. The latter, furthermore, hasten the progress of the suppuration, and as the eye is already doomed, this is usually the best thing which can be done. If the chemosis is very pronounced, so that it prevents the eyelids from closing over the eyeball, scarifications of the œdematous conjunctiva are sometimes indicated.

If seen in its early beginning, the trial should be made to combat the disease with all antiseptic measures at our disposal, including the attempt to remove any foreign body that may be lodged in the eye. Intraocular injections (*Abadie*), which have of late been highly recommended, are still *sub judice*, but may be tried. Of all the antiseptics used for intraocular injections, chlorine-water seems to be the best borne and most effectual.

Later on, when it is evident that all efforts to bring the disease to a standstill are futile, the choice may lie between evisceration and enucleation of the eyeball. The question is a mooted one, when such an eye should be enucleated, and it has been stated that to enucleate during the active stage is apt to produce cerebral sepsis and death. This reasoning seems decidedly faulty. The septic focus lies within the eye (except in metastatic choroiditis), and its removal under antiseptic precautions can prevent cerebral sepsis to which such an eye otherwise may give rise. It should, therefore, be removed as soon as it is evident that the disease cannot be arrested.

Gummata undoubtedly occur in the choroid, but they are seldom recognized as such.

Tubercles are sometimes found in the choroid during the tuberculosis of the lungs. In rare cases the occurrence of tubercular choroiditis is a primary affection. When the diagnosis is primary tubercular choroiditis, the immediate removal of the eyeball is indicated.

§84. The malignant tumors of the eyeball take their origin most frequently from the tissue of the choroid. These tumors are *sarcomata*, and are either pigmented or, in rare cases, unpigmented. They may, furthermore, contain newly-formed cartilage or bone-tissue.

The marked clinical symptoms presented in its different stages by such a growth within the eyeball ought to be known to every physician. In the first stage the patient is gradually losing sight, although externally the eye shows nothing pathological, except, perhaps, a few dilated bloodvessels in the episcleral tissue. In the rare cases in which an eye comes under observation at this stage, it may be possible to recognize the tumor with the ophthalmoscope. The retina may be simply lifted up by it, or may be detached at the site of the growth. In the second stage the growing tumor produces inflammatory symptoms which are plainly visible without the ophthalmoscope. The intraocular tension is increased, the crystalline lens and iris are pushed forwards, and the general picture of glaucoma (See Chapter XVII) is the result. There is usually considerable hyperæmia and pain during this stage. As it may yet be possible in the glaucomatous stage of the intraocular tumor to save the patient's life by the removal of the eyeball, the physician should be well posted on this symptom. (See Fig. 63). Finally, in the

FIG. 63.—Primary melanosarcoma of the choroid. The retina is lifted up by the growth.

third stage the eye is totally blind, the iris and cataractous lens are pushed forwards against the cornea and gradually the tumor breaks through the sclerotic, or possibly the cornea, and grows either into the orbital tissue or through the anterior portion of the eyeball out of the orbit. It may also grow along the course of the optic nerve. It may fill the whole orbit and grow out of it to a considerable size. Superficial ulcerations and hæmorrhages are then common.

During this last stage, as a rule, metastatic tumors have already begun to be formed in the brain, the lungs, the liver, or some other distant organ of vital importance, and the patient is irrevocably lost. It usually takes several years to reach this end.

There is no treatment for such tumors. The only thing to

be done is to remove the affected eyeball at the earliest possible period. Everything else is useless, and delay only enhances the danger to the patient's life. Yet, as there are but few people who are intelligent enough to submit at once to the inevitable, the family physician will generally be applied to for advice, and he will often have the disagreeable duty involving on him to indorse the statement of the oculist, and to impress the patient in the most earnest manner with the danger of his condition. The early removal of such an eyeball often saves the patient's life, and is surely the only means by which this result may be obtained. In some cases, however, relapses occur within the orbit, and metastases may have been formed in other parts, even if the eye was removed at so early a period that the hope seemed justified that dissemination had not yet taken place.

§85. Injuries to the eyeball by a blunt instrument which does not pierce the sclerotic or rupture it, may lead to *isolated ruptures of the choroid.* (See Fig. 64). After such an injury

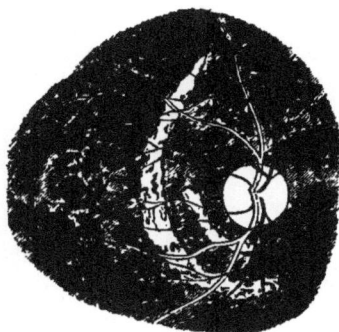

FIG. 64.—(After Knapp). Isolated rupture of choroid.

the fundus of the eye is usually at first obscured by blood and no details can be seen. When the blood is absorbed, the patients usually notice an impairment of sight, and sometimes complain of metamorphopsia, just as in a case of central choroido-retinitis. We find then, on ophthalmoscopic examination, a rent in the choroid forming a crescent, generally concentric with the optic disk and near it, through which the sclerotic is seen. The edges of the rent are pigmented and the untorn bloodvessels of the retina run across it.

Hæmorrhages are but seldom seen in the choroidal tissue, except after injuries and during some forms of choroiditis.

CHAPTER XIII.—DISEASES OF THE RETINA.

§86. *Hyperæmia* of the retinal bloodvessels is but seldom seen without being caused by some other eye affection. It is always combined with hyperæmia of the optic papilla. When retinal hyperæmia is the primary affection, it may produce no pathological changes in the structure of the retina; it may, however, produce hæmorrhages. It is usually caused by overwork, especially in bad light. Rest, and the moderate exclusion of light will generally suffice to do away with its symptoms, which consist chiefly in the weakness of sight. Quite often an error of refraction is the predisposing cause, and its correction by glasses will be followed by a return to the normal condition.

§87. *Anæmia* or *ischæmia* of the retinal bloodvessels, without accompanying pathological changes in the tissue of the retina, is sometimes seen as a sequel of a severe, prostrating illness, especially when the heart's action is considerably enfeebled. Such an ischæmia of the retina causes partial or total blindness, which may be momentarily relieved by puncture of the anterior chamber; the weakened heart may then suffice to overcome the reduced intraocular pressure, and to force the blood again into the small retinal bloodvessels.

A similar condition, but due to a contraction of the bloodvessel walls, may be caused by large doses of quinine. I have once seen it occur after a very severe fright. In this case the normal condition was restored after a very few days. The amblyopia, caused by anæmia dependent on prostrating disease, usually disappears as the patient recovers strength and health.

A partial or total anæmia of the bloodvessels of the retina, with subsequent anatomical changes in the retinal tissue, is, furthermore, observed in consequence of *embolism* of the *central retinal artery* or of one of its retinal branches, or in consequence of *thrombosis* of the *central retinal vein.* (See Fig. 65).

In embolism of the central retinal artery the partial or total blindness occurs suddenly and the ophthalmoscope reveals a perfect anæmia of the retinal arteries; the retinal veins are attenuated, but contain here and there broken colums of blood. Near the optic papilla and around the macula lutea, the retina appears obscured by a whitish infiltration, and the fovea centralis is seen as a small cherry-red spot. This pronounced color may be due to a hæmorrhage behind the fovea centralis, but is probably always due to contrast. After some time the infiltration in the retinal tissue disappears, and later on the retina and optic nerve become atrophic.

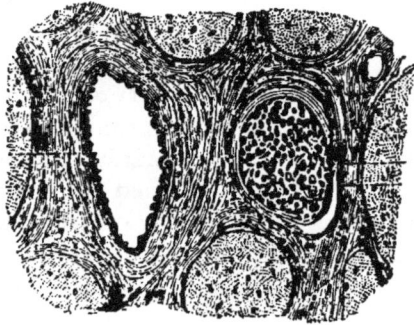

FIG. 65.—(After Michel). Thrombosis of the central retinal vein, six millimeters behind the entrance of the optic nerve.

In embolism of a single branch of the central retinal artery these symptoms and the loss of function are confined to the part of the retina which the affected branch supplies.

Embolism of the central retinal artery is usually due to some valvular disease of the heart. When seen early it may in exceptional cases be relieved by gentle massage or puncture of the cornea. By these means the embolus may become detached and be driven into a smaller and less important branch.

Septic embolism will cause purulent inflammation of the retina and other structures of the eye.

In thrombosis of the central retinal vein the arteries appear also very thin, but the veins are usually greatly enlarged, and a number of hæmorrhages are seen in the retinal tissue. It may also affect but one branch of the central vein. It is due to heart disease and senile changes in the walls of the bloodvessels.

When facial erysipelas causes blindness, as it sometimes does, it may be due to thrombosis of the central vein produced by the inflammation and swelling of the orbital tissue (*Knapp*).

§88. Another form of more or less sudden blindness owes its origin to the *detachment of the retina* from the choroid. This is due either to an effusion of a serous fluid between the two membranes, (and in this manner it is most apt to occur in eyes suffering from a high degree of myopia), or it follows the shrinkage of the vitreous body from some cause or other (loss during an operation or injury, fibrinous degeneration). (See Fig. 66).

FIG. 66.—Detached retina (after von Wecker and Jæger.)

Patients attacked by detachment of the retina usually state that at first a cloud seemed gradually to spread over the sight until it allowed them to see only a part of the object looked at, and this only as if through a mist; at a later stage sight is reduced to virtually nothing. It may have taken a few hours, or days, or even months, to develop all these symptoms.

When a large portion of the retina is detached, and especially when the detachment involves the peripheric parts of the retina, the detached bladder-like membrane may sometimes

be seen with the naked eye floating behind the crystalline
lens.

By the examination of the field of vision the extent of the
detachment may readily be judged; examination with the ophthalmoscope will enable us to make the diagnosis sure.

Sometimes we see small rents in the detached parts of the
retina, caused evidently by the pressure of the fluid exuded
behind it.   According to *Leber* and *Nordenson* these rents
precede the detachment and are due to the pulling of a shrinking vitreous body.

The vitreous body in the myopic form of detachment of the
retina is usually liquified.   The intraocular tension of the eyeball is less than normal.   (See Fig. 67).

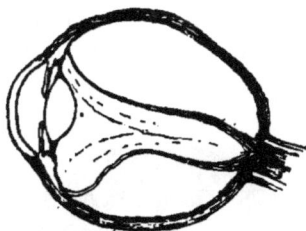

F IG. 67.—Total detachment of the retina.

The diagnosis is, as a rule, easily made.   The only affection
with which detachment of the retina might easily be confounded is sarcoma of the choroid, in which, moreover, retinal detachment is often actually present as a complication.

The prognosis is unfavorable.   Although we may sometimes
succeed in perfectly curing a recent detachment of the retina,
relapses are almost sure to occur, and getting less and less
tractable, they finally leave the eye in a useless condition.

The treatment now most practiced, consists in repeated hy-
podermic injections of a solution of the muriate of pilocarpine,
and the results, especially in recent cases, are often for the
time very satisfactory, and may remain so permanently.   This
method of treatment requires less time and is less disagreeable to the patient than others.

Among these, the most prominent one is prolonged enforced rest of the patient and of the eyes.

This is accomplished by keeping the patient in bed, and even as much as possible on his back, by paralyzing his accommodation with sulphate of atropia, and by slight, continued pressure with a compressive bandage. To withhold fluid, as much as possible from the patient, and even to give him a free allowance of salt with every meal, is helpful.

The operative treatment of detachment of the retina consists in puncturing the sclerotic and choroid behind the detached retina, or to pierce with the needle through the sclerotic, choroid and retina into the vitreous body. In this manner the exuded fluid may be withdrawn from the eye and even an adhesive inflammation be set up at the site of the puncture, by which the retina may be held in place. Of late, iridectomy has been recommended, but it does not seem a rational procedure. The results of these operative measures are as insecure as those of other forms of treatment.

Later on, the intraocular tension of an eye suffering from detached retina is considerably below par, and the lens becomes cataractous, and often a low-grade iritis is observed.

Detachment of the retina usually attacks but one eye at a time, and but very seldom is seen to befall the fellow eye at a later period.

§89. *Pigmentation* of the retina may, as has been stated, be the consequence of plastic choroiditis, and it then, as a rule, remains stationary. There is, however, also a progressive form of pigmentation of the retina, due to proliferation of the pigment epithelium which is called *retinitis pigmentosa* (*pigmentary retinitis*). (See Fig. 68). This affection gradually leads to atrophy of the retina through solidification of the bloodvessels by an inflammatory process in their lymph-sheaths and walls, and atrophy of the optic nerve. It terminates usually in blindness. It takes, however, a great many years before this final result is reached. As the disease begins at the periphery of the retina, and only gradually advances towards its center, peripheric vision is first destroyed. Thus the visual field becomes gradually narrower and narrower, the patients see as if looking through a tube, until finally their central vision is also lost.

A very characteristic symptom in this affection is that the patients see much less distinctly in the dark. This condition is known as *hemeralopsia* or night-blindness.

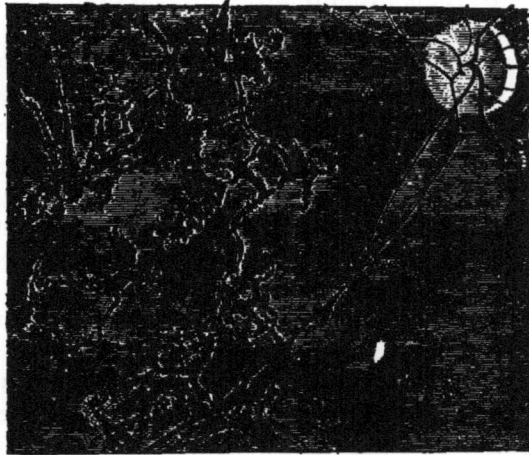

FIG. 68.—(After Liebreich).   Ophthalmoscopic appearance of the retina in retinitis pigmentosa.

Pigmentary retinitis is always observed in both eyes. Treatment is but rarely useful, although in its beginning the disease may sometimes be brought to a standstill, and in older cases vision may sometimes be improved.   Subcutaneous injections of strychnia, the constant current and, when there is a history of inherited syphilis, bichloride of mercury or some other mercurial should be tried.

With regard to its etiology, consanguinity of the parents has been thought by some writers to be the primary cause. Inherited syphilis may also be one of the causes.

The retinal tissue is not very apt to become primarily inflamed, and there are only two distinct forms of retinitis, not dependent on choroiditis, namely, syphilitic retinitis and albuminuric retinitis, the latter being actually a neuro-retinitis.

§90. *Syphilitic retinitis* is usually one of the later manifestations of syphilis.   The ophthalmoscope shows the retina dull and hazy, particularly in the vicinity of the optic papilla and along the course of the larger retinal bloodvessels. The bloodvessels themselves appear veiled, and sometimes perfectly covered in a part of their course by the exudation. The retinal veins are broad and somewhat tortuous. The optic papilla usually appears reddish and slightly swollen, and its outlines are not quite distinct. If the disease goes on, it leads to hypertrophy of the connective tissue of the retina, and consequent atrophy of its nervous elements. There is generally a dust-like exudation in the vitreous body.

The patients complain of seeing strange photopsiae, dazzling lights, etc. Their sight is obscured in such a way that they need more light than formerly, in order to see distinctly. The degree of impairment of vision is, however, often much less than would be expected from the condition of the retina.

Syphilitic retinitis is very prone to relapse, and if not properly treated, it may as already indicated, lead to blindness.

The treatment consists in vigorous anti-syphilitic measures, combined with perfect rest of the eyes in a dark room, assisted by the instillation of sulphate of atropia, and by local depletions.

§91. *Albuminuric retinitis* as its name indicates, is chiefly due to albuminuria *i. e.* to nephritis in all its forms, but especially to the shrinking kidney.

In an eye suffering from albuminuric retinitis we find the optic papilla swollen and infiltrated, its normal outlines indistinct or even perfectly hidden by a whitish exudation, which extends into the neighboring retinal tissue. The retinal veins are gorged with blood, and very tortuous, almost like corkscrews, their origin in the papilla is invisible, and parts of them within the retina are perfectly covered by a dense whitish exudation. The retina shows, moreover, a number of whitish, shining spots of various shapes. These appear, especially, around the macula lutea, and are usually arranged in a radial direction around it, thus forming a very characteristic stellate picture. (See Fig. 69). Here and there hæmorrhages are visible in the retinal tissue. In rare cases detachment of the

retina around the optic papilla has been observed.   Albuminuric retinitis is due to changes in the bloodvessels of the retina, and belongs to the uræmic stage of nephritis.   Yet, in
some cases, the ophthalmoscopic diagnosis may reveal nephritis, when its existence has not yet been thought of, because of
the absence of other marked symptoms of the disease.

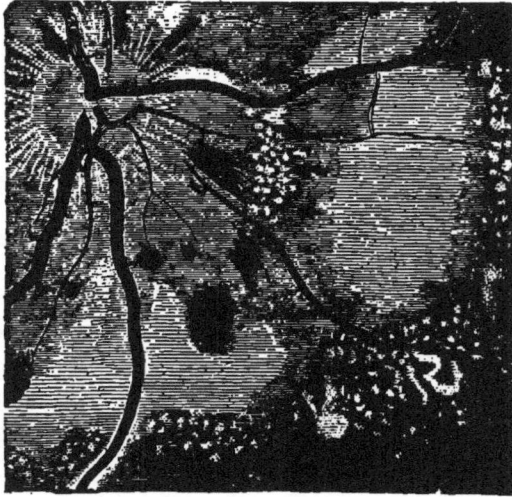

Fig. 69.—Ophthalmoscopic appearance of neuro-retinitis albuminurica.

The disturbance of vision is often strangely small, in view
of the very conspicuous changes in the structure of the retina;
on the other hand patients in later stages of the disease may
sometimes be blind for hours at a time and then regain sight,
without any visible changes in the condition of the retinal tissue. This happens during the so-called uræmic attacks (*uræmic
amaurosis*).

It is hardly necessary to say that, as there is no cure for a
well established nephritis, there is also none for the albuminuric retinitis.   The patients die generally within one, or at
most, two years after the retinitis has been first observed.

Similar forms of retinitis are sometimes seen in cases of intra-cranial tumors with subsequent neuro-retinitis. (See Chapter XXII).

§92. *Hæmorrhages* into the retinal tissue may be observed without being preceded by any inflammatory changes, and they may come on during inflammatory diseases of the retina. The retinitis is not due to the hæmorrhage and the old name of *hæmorrhagic retinitis*, is therefore, a misnomer. Their cause is a degeneration of the walls of the bloodvessels, or an embolism or a thrombosis. They vary in size, shape and number, may attack one eye alone, or both eyes at the same time, and accordingly they interfere with vision to a greatly varying degree. In some cases the retinal hæmorrhages are so numerous that sight is nearly abolished from the start.

The exuded blood soon undergoes fatty degeneration, and becomes absorbed, leaving an atrophic spot or spots, where the nervous elements of the retina have been destroyed. Sometimes, however, the absorption takes place, leaving no trace behind. In other cases a glaucomatous process may be set up leading to the destruction of the eye. Having once occurred in an eye such hæmorrhages are likely to recur.

The causes of the retinal hæmorrhages are heart disease, atheromatous degeneration of the walls of the arteries, malaria, pernicious anæmia, etc., or, in females, they may be due to the suppression of the menstrual flow, especially at the period of the change of life, and to thrombosis of the central retinal vein, or one of its branches.

The treatment of retinal hæmorrhages must adapt itself in the main to the particular causes which are recognized in the case in hand; at the same time it is well to give the eyes perfect rest, to give mild cathartics and perhaps, also, to apply leeches to the temple.

§93. The retina is sometimes the seat of a form of malignant tumor, which has been called *glioma*. It is only found in children and in many cases soon after birth. (See Fig. 70).

The affection is usually noticed by the parents through the child's blindness in one or possibly both eyes, together with a characteristic yellowish-gray reflection from the back-ground of the eyeball, which has given rise to the name of *amaurotic cat's eye.*

The tumor usually grows rather rapidly, and although the

eye may, so long as the tumor is small, show no external sign of anything being wrong within it, the newformation will at a later period cause exactly the same symptoms which we have described as due to the growth of a choroidal tumor. Finally the tumor bursts through the eyeball and shows itself externally. At the same time it has generally already invaded the tissue of the optic nerve. It soon causes metastases in the bones of the skull, the brain and other organs. The only remedy to save the patient's life is the earliest possible removal of the eyeball. When the disease has attacked both eyes at the same time the physician may find himself in the most disagreeable position of having to insist upon the speedy removal of both eyeballs in the hope of possibly saving the child's life. However, in such cases the parents usually refuse an operation and prefer to let the child die.

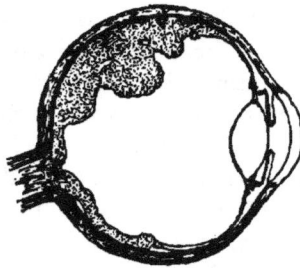

FIG. 70.—Glioma of the retina at a comparatively early stage.

When the eye is removed too late, and this, unfortunately, is nearly always the case, relapses will occur within the orbit, leading rapidly to death, although usually with less suffering to the child than when the eyeball has been allowed to burst and to form a huge and repulsive tumor.

In some cases inflammatory deposits in the vitreous body have been mistaken for glioma (*pseudo-glioma*).

CHAPTER XIV.—DISEASES OF THE OPTIC NERVE.

§94. *Hyperæmia of the optic papilla* is almost always accompanied by hyperæmia of the retina, and occurs as a symptom accompanying inflammations of the uveal tract. But we sometimes meet with a genuine hyperæmia of the optic nerve and retina alone, in which case it may be an early stage of a neuritis, or simply a passing pathological condition, brought about by over-work, especially in bad light. The symptoms and the treatment are the same as have been described in connection with hyperæmia of the retina. When the hyperæmia is caused by an uncorrected error of refraction, this should be corrected by properly selected glasses.

The same conditions which cause anæmia of the retina, produce also *anæmia* of the *optic papilla*. The anæmia may be due to an obstacle of the normal blood current or to a diminished blood supply from failure of the action of the heart. In the latter case it is chiefly due to excessive hæmorrhages, to weakness of the muscle of the heart from other causes, to anæmia of the brain, or to general anæmia.

§95. *Neuritis optica* appears in two forms: the more frequent one, in which the inflammatory symptoms of the optic papilla and its neighborhood are plainly visible (*papillitis(!), papillo-retinitis, neuro-retinitis, neuritis-ascendens*), and a second form in which the ophthalmoscope may at first reveal hardly any signs of inflammation, but yet the symptoms are such as can only be explained by the supposition that a neuritis is present in the part of the nerve which lies behind the eyeball (*neuritis descendens*).

We sometimes find, preceding neuritis ascendens, an œdematous condition of the optic papilla. This, however, is soon superseded by active inflammatory symptoms.

In most cases the neuritis involves first the instertitial connective tissue of the optic nerve, and the nerve-fibres become only secondarily affected.   (See Fig. 71).

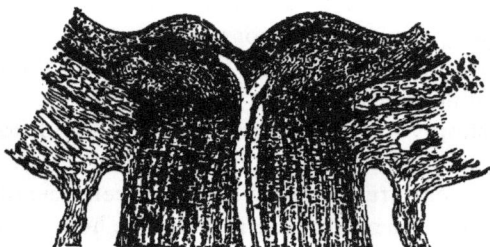

FIG. 71.—Longitudinal section through an optic nerve during interstitial optic neuritis.

On examination with the ophthalmoscope we find in early stages of optic neuritis a hyperæmia of the optic papilla, combined with swelling, and indistinctness of its outlines.   The retinal veins are enlarged and tortuous.   Later on the optic papilla becomes more and more infiltrated and swollen.   Its color is then a whitish-gray, its normal outlines become perfectly hidden, and small hæmorrhages may appear in its tissue. The main trunks of the retinal bloodvessels in the optic papilla become so covered by the infiltration that it may be impossible to recognize them.   The surrounding retinal tissue appears hazy, and hæmorrhages into the retinal tissue are seldom wanting.   (See Fig. 72).

The more the swelling and infiltration of the optic papilla increase, the more prominent are the symptoms of constriction of the retinal bloodvessels.   This fact has given this condition the name of "choked disk."   Later on the retraction of the newly formed connective tissue and consequent atrophy of the optic nerve and retina take place.   In these cases of atrophy after neuritis optica we are often able to diagnosticate the preceding neuritis long after it has occurred by the bluish-gray discoloration and irregular outlines of the optic papilla, and from the remaining traces of hæmorrhages.

If arrested in its early stages the disease sometimes gets well without impairment of vision.

The patients come generally under observation on account

of impaired vision; still, the original causes of the neuritis may have forced them to seek a physician's advice before the appearance of any eye-symptoms. The condition of the optic nerve may then help to diagnosticate the primary disease.

Fig. 72.—Ophthhlmoscopic appearance of the optic disk in optic neuritis, with striated hæmorrhages.

The commonest primary causes of this form of optic neuritis are intra-cranial affections, and among these especially tumors, meningitis, injuries to the skull, then albuminuria, syphilis, lead-poisoning and diabetes. The disease most frequently attacks both eyes.

The treatment of optic neuritis, as of retinitis, is in the main that of its primary cause.

We shall have further occasion to refer to this subject in Chapter XXII.

Neuritis descendens is seen more rarely.

The affection causes a more or less sudden total or partial blindness of one or both eyes, and with the ophthalmoscope we find at first, perhaps, only a somewhat anæmic optic papilla. This anæmia is caused by pressure upon the arterial bloodvessels. The retinal veins appear fuller than normal, and there may be also a slight exudation. Later on the optic papilla and retina may show signs of active inflammation.

The disease may end in perfect recovery or it may lead to a partial atrophy of the optic nerve and, especially, of those fibres which go to the macula lutea. In this case peripheral vision may be normal, while central vision is abolished (*central scotoma*).

If there is any general cause to be detected, to which the disease of the nerve may be due, treatment must be directed to this cause. When we are unable to find such a cause, the employment of subcutaneous injections of strychnia, iodide of potassium, mercury in some form, leeches, diaphoretics and diuretics may be beneficial.

FIG. 73.—Atrophy of the optic nerve and consequent flat excavation of the optic papilla.

§96. A frequent affection of the optic nerve is *atrophy* without any visible sign of inflammation, although histologically an interstitial neuritis has been found to be the rule (*Uhthoff*), and it is due undoubtedly to impaired nutrition and consequent degeneration of the nerve-fibres, produced by the slowly increasing newformation of connective-tissue. The atrophy may attack a part only of the optic nerve fibres, or it may involve the whole nerve.

The optic papilla in such a case appears whitish or grayish, sometimes dotted, in other cases of a uniformly shining white. The retinal bloodvessels are very small, and their peripheral ends and smaller branches are invisible. ͺ Later on the optic papilla appears slightly excavated. (See Fig. 73).

There are often central scotomata, and the color sense is diminished, or color-blindness, especially for red and green, may occur. According to the degree of atrophy is the impairment of sight; in the highest grades it is totally abolished.

The disease usually affects both eyes. A similar condition may come about from any pressure upon and consequent impairment of the nutrition of the optic nerve, as in cases of fracture implicating the optic canal (*Berlin*), or through an exostosis, or from exudation in cerebral or cerebro-spinal meningitis.

The most frequent causes of this form of atrophy of the optic nerve are syphilis, the abuse of tobacco and alcoholic stimulants, cerebral and spinal diseases. When the affection is due to chronic nicotine and alcohol poisoning it is usually preceded by a marked capillary hyperæmia of the optic papilla, the evidence of the interstitial inflammatory process mentioned above. Among the sufferers from this affection those engaged in the trade of alcoholic beverages are very frequent.

The atrophy of the optic nerve following the abuse of tobacco or of alcohol, or of both together, has a comparatively good prognosis, if the patient has enough moral energy left to abstain totally from the use of these noxious substances. Most patients of this kind, however, will resist the attempt to help them; and here the voice of the family physician will often be of great importance in aid of the efforts of the specialist. I have often heard the existence or even the possibility of such an affection laughed at, or at least doubted; yet, the fact is so absolutely certain, and the disease so perfectly recognizable and recognized, that every physician, who by his doubts aids the patient in evading the necessary restriction, commits a great wrong.

Besides enforcing abstinence from the indulgence in the use of tobacco in any form, or alcoholic beverages, it will be well to give aid to the usually poor appetite, to give iodide of potassium or some other tonic, and to induce sleep by bromide of potassium or hydrate of chloral. In most cases, however, strychnia is the best remedy, particularly when used in the form of subcutaneous injections and in comparatively large doses. Sometimes the use of electricity has proved successful. But all these remedies are of no use unless either total or, if this cannot be obtained, at least partial (especially with reference to the alcoholic beverages, as the general condition of some patients will not allow of a total withdrawal of all stimulants) abstinence from the use of the noxious substances is insured.

When syphilis is the cause of atrophy of the optic nerve anti-syphilitic treatment must be resorted to.

In a number of cases atrophy of the optic nerve is dependent on tabes dorsalis, or on sclerosis of the spinal cord or medulla and is then always associated with miosis and immobility of the contracted pupils, caused by the loss of reflex action. The pupil, however, usually becomes still smaller when an effort at convergence and accommodation is made (*Argyll-Robertson pupil*).

It has been already mentioned that in cases of atrophy color-blindness may occur and it must be considered a grave symptom.

This acquired color-blindness must, however, not be confounded with the not uncommon congenital color-blindness in which there is no disease, but a lack of perception of certain colors. The congenital color-blindness is usually either a red-green blindness, or a blue-yellow blindness. The acuteness of vision is generally perfect in this affection, and the color-blindness is incurable. (See Chapter XXIV).

The terms *amblyopia* and, especially, *amaurosis* are frequently used by practitioners, and seem to convey to the patients a most fearful idea of their condition. It may, therefore, be well to state that the term amblyopia means nothing but defective sight, and is, especially, used to designate defective sight from a disease of the back-ground of the eyeball, and more especially, of the optic nerve. Amaurosis, also, is not the name of a disease, but it means simply blindness from some disease of the fundus oculi, especially from an affection of the optic nerve.

The term *hemianopsia* designates the loss of half of the field of vision. It and its causes will be detailed in Chapter XXII.

§97. *Injuries* to the optic nerve are not of frequent occurrence; they are usually cuts or rents and tears by bullets or other blunt objects thrown into the orbit by an explosion or a similar accident. They lead invariably to partial or total atrophy of the optic nerve, according to the extent of the injury.

§98. The *tumors* of the optic nerve are either *myxomatous, fibro-myxomatous, endotheliomatous (psammoma)* or *sarcomatous* in nature. They usually spring from the sheaths of the optic nerve, and destroy sight, by compressing the nerve-tissue.

Aside from the gradually increasing blindness they cause also a gradually increasing exophthalmus. This protrusion of the eyeball is usually directed straight forward and, unless the tumor is very large, the mobility of the eyeball is, perhaps, impaired in a general way, but not in any especial direction. (See Fig. 74). This is easily understood, since the tumor originates within the hollow cone which the external ocular muscles form within the orbit. For the same reason we find by palpation that the tumor moves with the eyeball. The ophthalmoscope reveals either optic neuritis or atrophy of the optic papilla.

FIG. 74.—Myxoma of the optic nerve.

The treatment consists in an early removal of the tumor. Formerly the eyeball was always removed with the newformation, but since *Knapp* first succeeded in removing a tumor of the optic nerve without sacrificing the eyeball, the operation has been several times successfully repeated, and should always be attempted.

CHAPTER XV.—DISEASES OF THE CRYSTALLINE LENS.

The affections of the crystalline lens, for which patients seek advice, are of two kinds, namely, the formation of a cataract, or the dislocation of the crystalline lens. Both affections may be either congenital or acquired.

Cataract is the name for every opacity or dimness of the crystalline lens, be it partial or total.

§99. The *congenital* forms of *cataract*, with which we usually count those which are observed soon after birth, are: zonular cataract, polar cataract and total (soft or hard) cataract.

FIG. 75.—Two forms of congenital lamellar (zonular) cataract.

Zonular (lamellar) cataract is the name given to a condition in which one or more layers of the crystalline lens, or only parts of one or of several layers of its substance, are opaque. (See Fig. 75). It is usually stationary, seldom progressive, and nearly always affects both eyes. It is often connected with rhachitis or a strumous diathesis; it seems also to be in some cases connected with the fact of consanguinity of the parents.

The affection is usually first noticed by the parents several years after birth, and especially when the children begin to play with small objects. I have only once seen a case 24 hours after birth.

When the pupil is well dilated in such cases, we see behind it a greyish, circular, often striated, opacity, covered anteriorly by transparent lens-substance, and usually leaving the periphery of the crystalline lens clear. The opacity generally ap-

pears densest in the center, and its outlines are usually well
defined. If the outlines of the opacity are indistinct, it may be
a sign that the cataract is of a progressive form. By the use of
the ophthalmoscope it becomes evident what parts of the crystalline lens as yet remain clear and available for vision and, on
the other hand, we may judge whether some part of the lens is
in such a condition that it might, perhaps, be used to better advantage were the pupil more favorably situated. This point is
of great importance with regard to the method of operating
which is to be chosen. Eyes affected with a similar cataract
are often near-sighted, and may, furthermore, show other de-
fects, like microcornea, microphthalmus, or coloboma of the
iris, etc.

Zonular cataract usually prevents children from attending
school, and therefore something ought to be done to help them
at an early period. The operation which is frequently performed, and which is often quite successful, consists in making
a small iridectomy in front of the clearest part of the peripheral portion of the crystalline lens. An even better visual result
may be obtained by allowing the iris to remain entangled in the
corneal wound (*iridencleisis*) and this was at one time a legiti-
mate operation. By means of it the pupil will be dislocated
towards the side without the enlargement resulting from an
iridectomy, and some of the diffuse light, which otherwise
would fall into the eye, will be excluded. Although the immediate effect may be thus improved, the procedure of causing
an incarceration of the iris in the cornea is of somewhat
doubtful advantage, when we consider the future of such an
eyeball. In fact very serious consequences such as iridocyclitis, glaucoma, and even sympathetic ophthalmia of the other
eye, have been occasionally observed as a result of this particular procedure, which has therefore been abandoned. A simple small iridectomy or, when it can be safely accomplished,
a simple iridotomy, is therefore generally to be preferred.

A large proportion of the cases of zonular cataract receive,
however, but little benefit or even none at all, from any of
these operations. It is, therefore, necessary in a great number
of the cases to get rid of the whole lens by incising the anterior
lens-capsule and thus allowing the lens-substance to be slowly

dissolved by the action of the aqueous humor. This little operation, called discission, must generally be repeated several times until a clear pupil is obtained. The time thus consumed by the rather slow process of absorption may be considerably shortened by the extraction of part or all of the swollen lens-substance a few days after discission.

An eye which has been operated upon in this manner is, of course, deficient in refractive power and has also entirely lost the faculty of accommodation, yet with the help of proper glasses vision is usually comparatively good, and a moveable, round and central pupil is procured as in the normal eye. I have performed this operation quite a number of times with marked success in eyes even which had previously been operated upon by iridectomy with little or no benefit, and I personally prefer it to iridectomy in almost every case.

Anterior polar or *pyramidal cataract* consists in a small densly opaque cone, sitting apparently upon the middle of the anterior surface of the lens-capsule, but really enclosed within it.

What is sometimes called a *posterior polar cataract* is usually a deposit upon the posterior lens-capsule, and is, therefore, properly speaking, no cataract at all.

FIG. 76.—Congenital anterior polar (pyramidal) cataract and coloboma of the iris.

Anterior polar or pyramidal cataract is sometimes seen to develop, after perforation of the cornea in childhood, it is, however, usually a congenital affection, and sometimes repeats itself in several children of the same family. When congenital it may be the only visible defect or it may be combined with other malformations (microcornea, microphthalmus or coloboma of the iris, or of iris and choroid, etc.). (See Fig. 76). If no other malformation exists with it, it is not impossible that ulceration with perforation of the cornea and the subsequent adhesion of the anterior lens-capsule to the cornea, during fœtal life, have been the cause of the formation of the opaque cone. (*O. Becker*).

In an eye affected with this form of cataract the light can, of course, only enter peripherally. We may, therefore, be able to improve the sight materially by an iridectomy, or we may have to resort to discission of the lens-capsule in order to bring the absorption of the whole crystalline lens or to extraction of the lens, just as in lamellar cataract.

Total congenital *cataract* may be perfectly soft, or it may have a hard nucleus. Sometimes the whole cataract is thin and shrunken. In this case it has probably been, in the beginning, a soft cataract, and has lost some of its more fluid parts by exosmosis.

The removal of a total congenital cataract should be effected as early as possible, and the urgency is even greater than in cases of zonular or pyramidal cataract, inasmuch as the latter allows of some sight, while in total cataract there is no vision of objects, but only perception of light. The longer we delay the operation, therefore, the less will be the probability of obtaining good and useful vision. Soft total cataract may be removed by discission of the anterior lens-capsule or by simple extraction; harder ones had better always be extracted, and this, also, can mostly be done without an iridectomy. Shrunken cataracts may be divided or may be gently pulled out of the eye through a corneal section with forceps or a sharp hook.

In all three forms of congenital cataract *nystagmus*, i. e., a continued motion of the eyeballs, generally a pendulum motion, will develop, if the obstacle to distinct vision is not removed at an early date.

§100. More frequently than the different forms of congenital cataract we see those of *acquired cataract*. To designate the different stages of this process from its beginning to its end, we speak of an incipient, unripe (*immature*), ripe (*mature*) or over-ripe (*hypermature*) cataract. We call a cataract ripe when the lens is opaque throughout. This is determined by the fact that under oblique illumination the shadow of the pupillary edge of the iris is no longer visible in the lens-cortex and that we can no longer obtain any trace of a reflex from the back-ground of the eye with the ophthalmoscope. When the cataract is ripe the patient's sight is usually reduced to perception of light. In over-ripe cataract we find the outer or cortical parts fluid or semi-fluid, and we can often see the yellowish nucleus lying 'at the bottom of the lens-capsule, and changing its place, according to the law of gravitation. Such patients may then be able to see objects to a certain extent. Sometimes the total absorption of a cataract takes place and a good (aphakial) vision is restored.

With regard to the consistency of the cataract we have a soft or even fluid cataract, a semi-soft cortical cataract, a hard nuclear cataract.

The soft or fluid *cataract* is most frequently found in young people. It appears generally of a milky white or bluish color and the fluid contents seem to bulge the anterior lens-capsule forwards. In later stages deposits of lime or cholesterine may be seen in it, or the cataract may shrink by giving off some of its fluid parts. The anterior lens-capsule, which at an earlier period appears tense, will then become wrinkled, and it may in these cases, also, happen that, as a result of the shrinkage, a part of the pupil becomes uncovered and vision is more or less perfectly re-established.

Fig. 77.—Cortical Cataract.

Cortical cataract, the most frequent kind, consists in the primary loss of transparency of the cortical layers of the crystalline lens. (See Fig. 77). The formation of this form of cataract usually begins at the equator of the crystalline lens; and, in its incipient stages, it can be diagnosticated only after full dilatation of the pupil. It is then best seen by means of the ophthalmoscope. Gradually the whole cortex of the crystalline lens assumes a striated appearance, and the opalescent striæ are arranged around a center (the anterior pole of the lens) like the spokes of a wheel. As the disease advances these spokes become broader and denser and appear of a grayish-white or pearl-gray color. Finally, in the stage of ripeness, the whole cortex·is involved, and vision is reduced to the perception of light. While these changes are going on in the cortex of the lens, the nucleus generally retains nearly the normal degree of transparency and hardness belonging to the age of the patient.

Nuclear or *hard cataract,* is the senile cataract proper. In this form of cataract the central hardening of the crystalline lens, which in middle and advanced life results in the formation of the nucleus, goes on to the very periphery of that organ. (See Fig. 78). Such a cataract is hard, and has an amber tint. It does not, as a rule, entirely obscure sight, and it often allows patients to count fingers at some distance, even when it is perfectly ripe. This kind of cataract develops slowly, and the striæ in the cortical substance may be but little pronounced.

FIG. 78.—Nuclear Cataract.

When these different forms of cataract, which usually appear at a more advanced period of life, are seen in an otherwise apparently healthy eye, they are called *uncomplicated cataracts.* When other affections, or their results, are present in a cataractous eye, they are called *complicated cataracts.* Thus a cataract may be complicated by corneal scars, posterior synechiæ, atrophic choroiditis, detachment of the retina, glaucoma, synchisis of the vitreous body, atrophy of the optic nerve, etc. Complicated cataracts are generally much less favorable for operation than the uncomplicated ones, and the resulting vision is below par.

When cataract develops in young individuals its formation may be, and often is, due to *diabetes mellitus (diabetic cataract)*

The development of cataract in youth and in the early years of adult life is, moreover, often dependent on some inherited tendency, and may show itself in several members of the same family, like the congenital forms.

At a certain stage in the development of a cataract, the crystalline lens becomes swollen, while still tolerably transparent; and in this stage the eyes often become short-sighted, so that a patient who has long used glasses for reading, can read again without glasses, and may even see better at a distance with concave lenses.

The etiology of the formation of cataract is still shrouded in mystery. Nephritis has been claimed to be a frequent cause, but careful examination showed this idea to be fallacious. Of late some observers insist on its being due to a low degree of peripheral choroiditis, brought on especially by undue strain of the eye consequent upon uncorrected or badly corrected errors of refraction. *O. Becker* gave it as the result of his researches that cataract is an inflammatory disease of the crystalline lens, as shown by proliferation and metamorphosis of the capsular epithelial layer. If the theories of the inflammatory character, be it primarily in the choroid or in the capsular epithelial layer are correct, it is to be hoped that some means of treatment may be found by which to stay the process, and, indeed, some claims in this direction have been put forward.

Unfortunately in the vast majority of the cases of acquired cataract the dimness progresses in spite of all our efforts, and nothing remains to be done in order to restore vision but to remove the opaque crystalline lens from the eyeball, or at least from the axis of vision, by means of a surgical operation. The operation of discission, which has been already described, in connection with congenital cataract, is applicable also to softer acquired cataracts in persons under 30 years of age, or the division of the capsule may be followed, after the lapse of a few days, either by the removal of the soft and swollen lens-substance through a simple linear incision made for this purpose in the cornea, or by aspiration by means of a tubular curette. If the cataract is hard or has a hard nucleus, which is ordinarily the case in persons who have reached or passed the middle period of life, the operation known by the name of extraction, and which has for its object the removal of the opaque crystalline lens from the eyeball, is alone applicable. Extraction may be performed in several different ways, such as varying the form and position of the corneal incision, extracting the lens through the normal pupil, or enlarging the pupil by an iridectomy, opening the lens-capsule and allowing it to remain within the eyeball after the extraction of the lens-substance, or extracting the cataractous lens enclosed in and with the capsule, etc.

It does not fall within the scope of this book to give more than a bare sketch of such an operation as cataract-extraction, and for this the reader is referred to Chapter XXV. Every physician ought, however, to have a definite idea as to the best time at which to operate for cataract, and also to know something about the prognosis of such an operation.

Cataract extraction should not be performed until the cataract is ripe (and that is in most cases until vision is reduced to the bare faculty of differentiating between light and shade), except in special cases and when the patient is over 60 years of age. The signs by which the ripeness of a cataract is indicated have been detailed above. Unripe and over-ripe cataracts give often less satisfactory results after an operation than ripe ones. Before advising an operation, the physician must satisfy himself that the eye is in an otherwise healthy condition. It is, therefore, desirable that every physician should be able to examine the eye with regard to its function of vision. The patient should be able to see a candle-flame in an ordinary darkened room across the whole room, and also to point out in what direction the candle is held. If he cannot do this the cataract is usually in some way a complicated one. If there are no visible signs of former inflammation of the cornea or iris, the trouble must be sought for further back. By now examining the field of vision with a lighted candle in a darkened room, as detailed in Chapter II, we first of all confirm or correct the former observation made with the candle; and secondly, we examine whether there may be, perhaps, a well defined portion of the field of vision wanting, in which case the diagnosis of an atrophic choroiditis

or of a detachment of the retina is to be made, and the prognosis is modified accordingly.   The latter affection is the most
likely to be present if the eyeball is also soft.   If the pupil dilates but slowly, and then not fully even after the instillation
of sulphate of atropia, there is probably something wrong in
the uveal tract, and such eyes are less favorable subjects for
extraction of the cataract, even when the functional examination reveals no further disturbance.

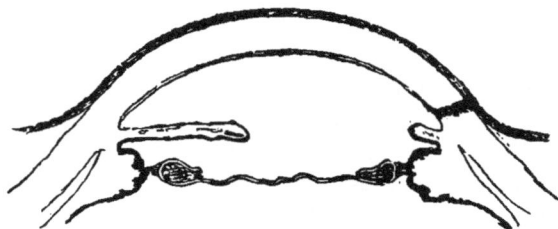

FIG. 79.—Secondary (capsular) cataract after extraction of lens by Graefe's method.

After cataract extraction has been performed by any of the
methods in which the lens-capsule is left in the eyeball, this
membrane may give occasion for a secondary operation
through its becoming dim and wrinkled, forming what is called
a secondary cataract. (See Fig. 79).   In such a case the
tearing apart of the lens-capsule (discission) as soon as may
be after the primary operation, is sufficient to improve vision
materially.   The longer we wait, the tougher the lens-capsule
will become, and it may finally become almost impossible to
cut it.   When there has been an iritis, and perhaps an irido-
cyclitis after the operation, the lens-capsule and the newly
formed membrane due to the inflammation, become glued together into one continuous membrane.   Iridotomy (iritomy)
may then be performed with the result of giving the patient
some useful sight, especially if the tension of the eyeball is
not diminished.

§101. *Injuries* to the crystalline lens by which the lens-
capsule is cut or ruptured cause the formation of cataract
(*traumatic cataract*).   When the wound of the capsule is very
small it may be so quickly closed by proliferation of the cap-

sular epithelium, that no appreciable quantity of aqueous humor comes in contact with the lens-substance, or the iris may
be driven into the capsular wound and thus plug it. If this
happens, the dimness of the crystalline lens may remain confined to the wound and its immediate neighborhood and never
progress. These cases are, however, rare exceptions. The
rule is that the capsular wound admits of sufficient imbibition
of the lens-substance with aqueous humor to render a large
portion, if not all, of it dim. Inflammatory reaction of the
capsular epithelium also helps in the progress of this form of
cataract.

FIG. 80. --Dislocation of a semi-transparent crystalline lens which is reduced in size
into the anterior chamber.

By a blow upon the eye with a blunt instrument the crystalline lens may be torn from its suspensory ligament (*zonule of
Zinn*), or the latter may be torn from the ciliary body. Thus
the crystalline lens becomes *dislocated.* The dislocation may
take place backwards into the vitreous body or forwards into
the anterior chamber (See Fig. 80), or the dislocated lens may
remain partly in the vitreous body and partly in the anterior
chamber. If the upper part of the zonule of *Zinn* is torn to
a small extent only, the crystalline lens may sink behind the

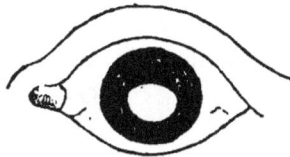

FIG. 82.--Dislocation of the crystalline lens into the vitreous chamber downwards,
showing the characteristic semilunar space of the pupil.

iris and a part of the pupil be thus freed from it. (See Fig.
81). If after such an injury the iris trembles (*iridodonesis*)

with the movements of the eyeball, the diagnosis of a dislocation of the crystalline lens is almost certain.   The periphery
of a dislocated lens, as far as it lies in the field of the pupil,
may be seen as a black arc, when viewed by diffuse light or
with the ophthalmoscope, in consequence of the total reflection
of the light; but as a shining, yellowish-white arc, when viewed
under oblique illumination.   For better demonstration colored
light may be thrown into the eye, when the arc will appear
colored by it.   When the whole of the crystalline lens is dislocated into the anterior chamber, it may fill the chamber so
completely that there may be some difficulty in recognizing
it until it becomes dim.   It usually causes an increase of the
intraocular tension with all its disagreeable consequences and
by these means the diagnosis soon becomes easy.

Dislocation of the crystalline lens (generally downwards)
may also happen spontaneously in consequence of synchisis
of the vitreous body and consequent weakening of the fibres
of the suspensory ligament.   This is the rule when dislocation
occurs in short-sighted eyes.   Dislocation of the crystalline
lens may be also a congenital defect, and is known by the
name of *ectopia lentis.*

An eye with a dislocated lens is, as far as the pupil is freed
from the lens, in the same condition as an eye which has been
operated upon by the removal of a cataract.   According to
its former refraction it may be emmetropic or far-sighted. The
part of the pupil behind which the dislocated lens lies is usually myopic.   An eye with a dislocated crystalline lens cannot
accommodate.

Dislocated crystalline lenses, as a rule, gradually become
dim. Their position and motility with the consequent mechanical irritation may, furthermore, lead to inflammation of the
iris, ciliary body and choroid.   A crystalline lens dislocated
into the anterior chamber generally causes glaucoma, as already
stated.   A partially dislocated crystalline lens need not be interfered with until it becomes cataractous or at least until its
presence causes further trouble in the eye, and the extraction
of such lenses is usually quite difficult.   The use of eserine or
any other miotic agent may, at first, by the contraction of the
pupil, be very agreeable to the patient, but its use cannot well

be kept up *ad infinitum* and it has, of course, no curative effect.

A crystalline lens dislocated totally into the anterior chamber should be at once removed by a peripheral section, as in an ordinary cataract extraction.

The absence of the crystalline lens, whether as a result of dislocation or of an operation for cataract, is designated by the name of *aphakia.*

The condition of aphakia brings with it of necessity the loss of the power of accommodation and a change in the refractive state of the eye. These conditions generally render the wearing of strong convex glasses necessary for both distant and near vision. In cases of myopia only, when of a high degree, the removal of the lens may render the eye emmetropic or at least so nearly so that such patients may be able to do without glasses for distant vision with comfort.

CHAPTER XVI.—DISEASES OF THE VITREOUS BODY.

§102. Affections of the vitreous body are always secondary affections and are due to the diseases of the membranes surrounding it, in most cases to diseases of the choroid and ciliary body. We distinguish between a *serous hyalitis* (*synchisis corporis vitrei*), a *fibrinous* or *plastic hyalitis,* and a *purulent hyalitis.* Yet, it must be thereby understood, that these forms of inflammation of the vitreous body (hyalitis) are only extensions of inflammatory processes of a corresponding character in the uveal tract, or in the optic nerve and retina. It often happens, however, that the changes in the vitreous body due to the hyalitis, continue long after the primary affection has run its course, or may even remain permanent.

We find this to be the case especially with the various forms of opacities observed in the vitreous body after a fibrino-plastic exudation into it.

The most common forms of opacities in the vitreous body are the so-called *muscæ volitantes* (*mouches volantes*). A patient suffering from such muscæ volitantes, when he looks at a bright surface, sees small black and gray dots and threads on or in front of it, which seem to float and to "run away" whenever he tries to "look directly at them." These dots and threads often present the appearance of strings of beads, and also take on other quaint shapes. What the patient really observes are the shadows cast upon the retina by fibrinous threads or by small aggregations of cellular elements in the vitreous body.

The presence of such minute cellular elements and fine threads may be demonstrated in the normal eye; therefore, muscæ volitantes are not necessarily to be interpreted as a pathological symptom. When a person has once detected their existence he is apt to look for them again, and then he finds to his dismay that they apparently increase in number, although in fact he simply sees now opacities to which he had previously paid no attention. This discovery may bring him to the physician.

The sudden appearance of muscæ volitantes in an eye, or the actual and rapid increase in their number, where they have been present and noticed before, is to be looked upon, especially in short-sighted persons, as an indication of a disturbance in the uveal tract, and, therefore, as an important symptom.

With this exception, muscæ volitantes, as a rule, call for no treatment, and, if we once succeed in convincing the patient of the fact that, although possibly annoying, they are really of no importance, his anxiety is relieved, and they usually cease to be troublesome.

Opacities in the vitreous body, when they are extensive enough to be detected by the ophthalmoscope, whether as conspicuous films or flocks, or as denser membranes, or as a general diffused muddiness, are of far greater importance.

In such cases the opacities materially interfere with sight, and they are moreover, the evidence of the actual or former existence of a fibrino-plastic choroiditis. They are often found in highly myopic eyes. The vitreous body is then usually fluid, and the dense shadows fly about with every movement of the eyeball.

Even when the choroiditis, which has given rise to the formation of such opacities, is cured, we may try to bring about their absorption. Mercury, decoctum *Zittmannii*, iodide of potassium, the constant current, and the muriate of pilocarpine, are useful remedies in such cases. Subconjunctival injections of mercury have of late been advocated (*Darier*). In other cases large membranous opacities have been torn with success by means of a small cutting instrument.

Newformations of large masses of connective tissue within the vitreous body are rare, but when present they seriously interfere with sight. They lie usually near the optic nerve entrance, and are due to inflammation of the optic nerve or retina (*retinitis proliferans*).

Synchisis scintillans is the name given to a liquified condition of the vitreous body in which crystals (mostly of cholesterine) have been formed. When these fly about they sparkle in the light like particles of silver or gold. We know of no treatment for this condition.

Hæmorrhages into the vitreous body (*hæmophthalmus*) may occur spontaneously from retinal bloodvessels, or are observed to accompany injuries to the ciliary body, the choroid or the retina. We shall speak further about them in Chapter XVIII. If the eye in such a case is examined with the ophthalmoscope, it may at first be impossible to get any reflex from the back-ground, or a darker or lighter shade of red may be seen according to the quantity of blood effused into the vitreous body. Such hæmorrhages may become perfectly absorbed, or the fibrine may remain and form floating or membranous opacities. If the hæmorrhage is very extensive, the eyeball will probably become atrophied.

The conditions caused by the presence of a foreign body in the vitreous will be detailed in Chapter XVIII.

§103. *Glaucoma (ophthalmia arthritica, choroiditis serosa)* is an affection of the eye the nature of which is, as yet, not perfectly understood. Its destructiveness and the insiduous slow progress of some of its varieties, make it a disease of very grave importance. The general practitioner should be perfectly familiar with its chief symptoms, the more so since they are of such a nature as to bring the patient to seek aid from the physician quite as often as from the oculist.

The name glaucoma is as dark as the disease, and in the present state of our knowledge has little meaning. It used to be said that glaucomatous eyes have a greenish pupil, and for this reason the name has been given to the disease. The greenish pupil is, however, but rarely seen. The name, therefore, is a misnomer.

The cardinal symptoms of glaucoma are, an *increase* of the *intraocular tension,* which renders the eyeball harder than it is in its normal condition, and the *loss of vision* which is chiefly (especially in the later stages), but not solely, due to the *excavation* and subsequent *atrophy* of the *optic papilla.* The increased intraocular tension must, of necessity, even before real atrophy of the nerve has begun, interfere with the blood supply to the retina, and will, if strong enough, bring about an interference with the ductile faculty of the nerve-fibres in this membrane. According to the degree of increase of intraocular tension we designate it as $+T$ ($+T_1$, $+T_2$, $+T_3$). (See Chapter IV).

Around the two cardinal symptoms, viz.: increased tension and loss of vision, other symptoms may be grouped which characterize the different stages and forms of the disease. With regard to the stages we differentiate between the *prodromal* stage, *glaucoma evolutum,* and *glaucoma absolutum.* When glaucoma attacks an eye without being preceded by some other disease likely to produce it, we speak of *primary* glaucoma; when the reverse is the case we speak of *secondary* glaucoma.

§104. The *prodromal* symptoms must be considered really as mild attacks of glaucoma. They may be little noticed by the patient or may alarm him sufficiently to seek aid. This stage of the disease may cover a period of months and even years. The early recognition of this condition is the more important as the disease at this stage is usually much more tractable than at a later one. As prodromal symptoms the patients observe recurring attacks of dimness of vision of varying degree. Rainbow colored rings may at the same time be seen around the gas or candle flame, similar to those appearing to a normal eye when viewing a light through a window bespattered with minute rain-drops in a misty atmosphere. The range of accommodation is reduced, and in consequence presbyopia is developed at an abnormally early period of life, or an existing presbyopia rapidly increases. In some cases myopia develops when the whole of the crystalline lens is pushed forward and the fibres of the suspensory ligament are stretched. There may be pain connected with the attacks of dimness of vision which varies considerably. Sometimes it is but slight, in other cases it is excruciating and extends from the eye to the neighborhood, and even to the back of the head. It may also lead to nausea and vomiting. As soon as vision is permanently reduced, the glaucoma has passed from the prodromal stage to that of glaucoma evolutum.

The forms in which glaucoma evolutum is most frequently observed are *simple chronic glaucoma, acute inflammatory glaucoma,* and *chronic inflammatory glaucoma.*

§105. In *simple chronic glaucoma* the eye usually shows no external signs of disease. The chief symptoms are the loss of visual acuteness, the (sometimes imperceptible) increase of intraocular tension and the excavation of the optic papilla. The so-called *physiological* excavation of the optic papilla has no sharp edge and never reaches close to the periphery of the papilla. A typical glaucomatous excavation has a sharp edge, reaches very close to the periphery of the papilla and is often much deeper than the physiological excavation. When a glaucomatous disk is viewed with the ophthalmoscope the arrangement of the bloodvessels is very striking. There is usually a yellowish ring around the periphery of the papilla due to the stretching and consequent atrophy of the choroid. When the bloodvessels have passed this ring in the direction towards the papilla, they bent abruptly over the edge of the disk disappear partly or totally (See Fig. 82) and reappear in a parallactic deflection at the bottom of the excavation. If the excavation is deep, the parts adjoining its rim and those lying in the depth of it, cannot be seen plainly at one and the same time. By this means the depth of a glaucomatous excavation may be measured with the ophthalmoscope.

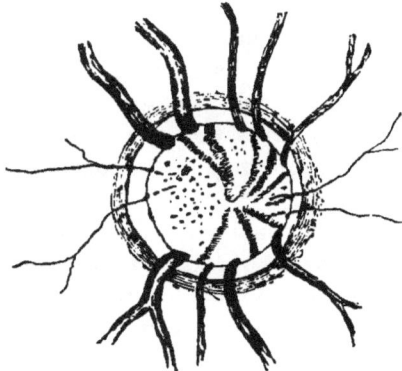

Fig. 82.—Ophthalmoscopic appearance of a glaucomatous excavation. The atrophied nerve fibres allow the lamina cribrosa to be seen as dark points.

At the bottom of the excavation the optic nerve appears punctated. This is due to the atrophy of the nerve-fibres which allow the net-work of the lamina cribrosa to be seen. There is often a spontaneous pulsation of the arteries, or this symptom may be readily produced by pressure on the eyeball.

As the visual acuteness becomes gradually lessened by the growing excavation of the papilla and atrophy of the nerve fibres, central vision (central scotoma), as well as peripheral vision become affected. The *contraction* of the *visual field* begins most

ly on the nasal side and extends gradually upward and down
ward. Finally, the visual field may be represented by a small
elliptical space, with the fixation point near its nasal focus. During the progress of these symptoms the *light sense* becomes
also diminished.   Color perception remains good.

§106.   In *acute inflammatory glaucoma* the disease attacks
the eye suddenly, oftenest at night.   It makes itself known
particularly by excruciating pain in the eye and head, sometimes combined with nausea or vomiting.   There is usually a
hypersecretion of tears and watery discharge from the nose,
and œdema of the upper eyelid.   The admission of light into
the eye causes the pain to increase.   The conjunctival and
episcleral bloodvessels are found hyperæmic, as in severe iritis,
the veins are particularly tortuous.   The cornea is steamy and
almost anæsthetic to the touch.   The eyeball is hard.   The
anterior chamber is shallow, its contents are muddy, the pupil
is wider than normal and acts sluggishly, or not at all.   No
details of the back-ground of the eye can be seen, often not
even a red reflex.   Vision may be reduced to the bare perception of light.

Such an attack may last a few hours or cover several weeks.
When the attack has passed off, vision may again be quite good.
Other acute attacks may follow, or this form may pass over
into the chronic form of inflammatory glaucoma.

In a few rare cases one such attack is known to have destroyed vision absolutely (*glaucoma fulminans*).

§107.   *Chronic inflammatory glaucoma* is the form in which
the disease is most frequently observed.   It may follow an
acute attack or develop slowly.   The external symptoms of
inflammation are less marked but are all visible.   The cornea
may be clear or hazy.   The anterior chamber is swollen, the
pupil wide and acts sluggishly or is immoveable.   The iris becomes discolored, its tissue atrophies.   The intraocular tension is increased.   The optic papilla is excavated.   There are
attacks of pain especially at night, and vision is gradually destroyed.

When by any of these forms of glaucoma sight has become

abolished, we speak of *glaucoma absolutum.* The disease, however, progresses still further. The eye remains painful. The lens is pressed forward till it touches the cornea and becomes cataractous (*glaucomatous cataract*). The cornea may become ulcerated and even perforated, or scleral staphyloma may develop. In short, the patient gets no relief until the eye is removed.

§108. *Secondary glaucoma* follows various diseases of the eye. The most frequent among these are: seclusion and occlusion of the pupil by plastic iritis, serous iritis, serous choroiditis, hæmorrhages into the tissue of the retina (probably due to the same cause as the seemingly secondary glaucoma) (*glaucoma hæmorrhagicum*), rapid swelling of the lens-substance when the lens-capsule has been ruptured intentionally or by an accident, dislocation of the lens, especially into the anterior chamber, scleral staphyloma, intra-ocular tumors, and some forms of keratitis (vesicular and ribbon-shaped keratitis).

§109. Glaucoma is observed in about one per cent. of the eye patients, and is more frequent among females than among males. It rarely attacks patients under fifty years of age. It may affect one eye only at a time, but the fellow eye almost invariably falls a victim to the disease at some later period. Sometimes an operation on the first affected eye seems to hasten the attack in the fellow eye. Eyes that are hypermetropic or astigmatic are more prone to be afflicted with glaucoma. In rare cases it has been seen to attack myopic eyes.

It may be well to state here that when the symptoms complained of by a patient point at all to glaucoma, the physician should not instill sulphate of atropia, hydrobromate of homatropine, or even cocaine into the patient's eye, as mydriatics are apt to bring about an acute attack.

§110. The anatomical changes found in glaucomatous eyes are grouped particularly around the optic papilla and nerve, and the ciliary body and the iris-angle.

The most striking, invariably characteristic histological change is the excavation of the optic papilla and atrophy of the optic nerve fibres. (See Fig 83). In a longitudinal section through these parts it is seen that the bottom of the excavation lies behind the level of the choroid. The meshes of the lamina cribrosa are closely pressed together and some-

FIG. 83.—Very deep glaucomatous excavation of the optic papilla. The nerve fibres are pressed aside and atrophied. The lamina cribrosa is pressed out of sclerotic. The optic nerve is atrophied.

times this portion of the sclerotic forms a convex line with its convexity reaching beyond the posterior surface of the surrounding sclerotic. A few nerve-fibres are seen lying in front of the lamina cribrosa, and lining the walls of the excavation and then to join the retina. Few bloodvessels are found and the veins are considerably wider than in the normal. At the edge of the excavation the retina and choroid are drawn down into it and towards its axis and are atrophic. The choroidal bloodvessels around the optic papilla are mostly obliterated and the pigment granules of destroyed cells lie free in the neighboring sclerotic. Sometimes small nests of infiltration are found in the choroid near the excavation. The optic nerve farther back is atrophied and the connective tissue trabeculæ are very much thicker than normal.

The dilatation and rigidity of the pupil are explained by the *adhesion* of the periphery of the *iris to* the posterior sur-

face of the cornea (See Fig. 84), due to the proliferation
of Descemet's endothelium and the new formation of a

FIG. 84.—Adhesion between the periphery of the iris and cornea as observed in
glaucoma.

small quantity of connective tissue.   In this manner *Fontana's*
spaces are obliterated.   In recent cases round cell infiltration

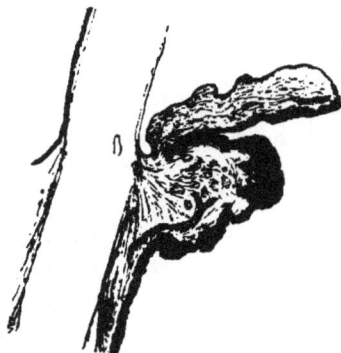

FIG. 85.—Swollen and hyperæmic ciliary body in glaucoma.

is found in the corneo-scleral tissue. *Schlemm's* canal is sometimes pervious, sometimes closed. This adhesion of the periphery of the iris to the cornea may be due to the general increase of the intraocular pressure, or to direct pressure on the iris
and lens by the swollen and hyperæmic ciliary body (*Brailey,
Weber*). (See Fig. 85).   Later on the ciliary body and iris are

found to be atrophied and many of the bloodvessels are obliterated.   (See Fig. 86).

FIG. 86.—Atrophic iris and ciliary body as they appear in the later stages of glaucoma. Schlemm's canal is obliterated.

These are in the main the pathological changes found in
glaucomatous eyes, and they but partially explain the symptoms of this still mysterious disease.

To unite and explain all symptoms of glaucoma by one common cause has been the aim of a large number of theoretic
endeavors.   Those that probably come most near to the truth
see this cause for all the symptoms in a disproportion between
the quantity of fluid secreted into the interior of the eyeball
and the amount excreted from it.   This may come about by
an abnormal increase of secretion (*Graefe, Arlt, Sattler*), by a
venous stasis (*Stellwag, Mauthner*), or by the obstruction of
the natural channels of drainage without any hypersecretion
(*Knies, Ulrich, Priestley-Smith, Brailey, Weber*).   *Schœn* more
recently sees the cause of glaucoma in an overstrain of the accommodation, in hypermetropic and astigmatic eyes especially,
and in its mechanical influence upon the structures of the eyeball.   Some authors think the excavation is chiefly due to
changes in the minute bloodvessels (*Haller's ring*) which lie
in the sclerotic arund the optic nerve entrance (*Jæger, Schna-
bel, Klein*).   *Risley* most recently tried to establish on a better
basis and with more strength an older theory, that glaucoma
is due to gout, by drawing attention to the similarity between

an acute attack of gout with one of glaucoma (*ophthalmia arthritica*).

All of these theories, and a number not mentioned here, seem to give some true explanation of one or the other group of symptoms. Yet, not one covers the ground fully.

§112. Unless interrupted by treatment in its progress glaucoma always leads to blindness. This treatment may be medical or surgical, and is the more successful the earlier the stage of the disease. As is well known, the miotic drugs not only contract the pupil but also reduce the intraocular pressure (physostigmine, sulphate of eserine, muriate of pilocarpine). The instillation of one of these drugs may act very benificially in the prodromal stage or may even cut short mild acute attacks, yet the effect is not lasting. As a rule, the progress of the disease can only be brought to an end by surgical means. The chief of these is iridectomy. It has become fashionable to divest great geniuses of their glory, but all that has been said to the contrary cannot rob *Von Graefe* of the laurels, won by teaching the fact that an iridectomy will heal a glaucoma. The knowledge that in a number of cases it is not followed by success, and in a smaller number seems even to hasten the destruction of a glaucomatous eye, should not induce us to withhold from the majority of the patients the benefit of this operation. Its value is undoubted in acute and chronic inflammatory glaucoma, while in simple chronic glaucoma it is least successful. The manner in which an iridectomy will lastingly reduce the intraocular pressure is not fully understood.

In stead of iridectomy, *sclerotomy* has been introduced by *Quaglino, Wecker* and others. It has proven to be almost useless in the inflammatory forms of glaucoma, while in simple chronic glaucoma, or after an iridectomy has not proven fully successful, it may be resorted to. Other operative measures, like *Hancock's* cut through the ciliary muscle, or tenotomy of the ciliary muscle, are hardly worth mention.

CHAPTER XVIII.—INJURIES OF THE EYEBALL
AND THEIR CONSEQUENCES.

§113. *Injuries to the eyeball*, especially when the foreign body which has caused the injury remains lodged in the eye, are of the gravest importance. Their prognosis, except when they are superficial only, must always be a doubtful one at first, as we have no means by which to judge whether the offending substance was aseptic or whether at the very moment the injury was inflicted, septic matter was carried into and became lodged within the tissues of the eye. Even when such an infection did not take place at the time the injury happened, the contents of the conjunctival sack may have produced a secondary infection when the patient is seen or produce it afterwards.

It is our duty, therefore, whenever we have to deal with an injury to try at once by all means to render the eyeball and the conjunctival sack aseptic, and to keep it so, as much as possible, until the injury is healed.

When septic infection has taken place and its results are already evident when we first see the patient, our means of combatting it, except in superficial wounds, are but small, indeed.

As the general practitioner is very often the first one to see such cases, he should always, before doing anything else, try to render the wounded eye aseptic. This is accomplished by carefully removing all foreign matter and flushing the conjunctival sack repeatedly with a solution of bichloride of mercury (1 in 3 to 5,000), or a four per cent. solution of boracic acid, or by dusting iodoform or aristol into the conjunctival sack or by staining the tissue with pyoktanine. Whichever remedy is used it must be used in sufficient quantity, and the rule should be to use rather too much than too little. If necessary atropine must be instilled. Immediately after the use of the antiseptic and when a condition of comparative asepsis has been produced, the eye must be closed, and this is best done in the manner described in Chapter VI, by absorbent cotton saturated with bichloride of mercury and adhesive plaster.   In this manner we may succeed in keeping the eye sufficiently aseptic to prevent infection of the wound, if it has not yet taken place, and to render an established superficial infection harmless.   If there is any obstruction to the tear-drainage and stagnation of the tear-fluid in the lachrymal sack, this must at once be attended to, as the lachrymal sack is under such circumstances apt to contain a great deal of infectious material.   If a superficial wound shows the results of an infection by the formation of pus and necrosis, there is no remedy higher to be prized than the immediate destruction of the infected portion by galvano-cautery or the actual cautery.   If these are not at hand, cauterization with nitrate of silver or pure carbolic acid may take the place.

If the infection concerns the tissue in the depth of the anterior chamber, opening the chamber and washing it out with a warm boracic acid or weak sublimate solution is often still valuable.   If the infective material has been deposited in the portions of the eyeball lying farther backwards, subconjunctival injections of a solution of bichloride of mercury or even intra-ocular injections of chlorine water should be tried. Atropine should be instilled in these cases.

§114. In the foregoing we contemplated injuries without the retention of a foreign body (except the infectious material) within the tissues of the eye. The rules, here given, hold good, even if a foreign body is retained.   To these measures, however, must then be added the removal of the foreign body, where it is possible.   The manner in which this has to be accomplished will be detailed farther on.   Aside from the danger of infection, all injuries to the eye are apt to influence unfavorably its further usefulness as an organ of vision in some measure.

Simple cuts in the cornea, which do not penetrate its whole thickness, heal, as a rule, without trouble and without interference.   When they are originally or become later on infected they lead to local infiltration, and the formation of a necrotic ulcer with all its consequences.   Every cut necessarily leaves a scar, which according to its size and situation, will interfere more or less with sight.

If a cut has penetrated the whole thickness of the cornea, the aqueous humor escapes and the anterior chamber is emptied. . When this happens, the contents of the posterior part of the eyeball are pressed forward, so that the iris, or crystalline lens, comes in contact with the inner opening of the corneal wound and plugs it, or becomes caught between its lips (*incarceration*)

FIG. 87.—A fold of iris-tissue is incarcerated in a corneal scar.

(See Fig. 87). In other cases the iris is carried through the wound canal to the surface of the cornea and beyond, and is held in

FIG. 88.—Recent prolapse of the iris through a corneal wound. The epithelium of the cornea is beginning to grow over the prolapsed iris.

this position (*prolapse*) (See Fig. 88). This prolapse may give rise to further troubles. The portion of the iris which lies outside the cornea is, as a rule, gradually cast off or shrinks, since by the constriction it is deprived of its nutrition. In other cases, when the nutrition is not cut off, it may simply become covered with corneal epithelium or it becomes the starting point of a granuloma, and then it grows as far as the eyelids will allow it. (See Fig. 89). Prolapse as well as incar-

FIG. 89.—A prolapse of the iris due to an injury has given rise to the formation of a granuloma (traumatic granuloma of the iris).

ceration of the iris, produce what is called anterior synechia of the iris. The pupil in these cases is drawn towards the scar which usually gives it a pear-shaped appearance, and every contraction of the sphincter pupillæ muscle pulls at this false insertion of the iris. This may give rise to a chronic state of irritation in this membrane, and ultimately to serious disturbances, such as iritis and, perhaps, glaucoma. We should, therefore, in the beginning, do what is in our power to restore the iris to its normal position within the eye. Gently rubbing the cornea with the eyelids, or prying the wound-lips apart with a thin curette, or directly pushing the iris into its normal position by means of a spatula, may sometimes suffice not only to liberate the incarcerated iris, but to keep it also from again prolapsing. If the injury is very recent, the action of a miotic helps in bringing about the desired effect. If the iris remains prolapsed, it is best to cut off the prolapsed part, and free the remaining iris altogether from its attachment to the corneal wound. This will virtually amount to making an iridectomy, and can be done even a few days after the prolapse of the iris has occurred.

In some cases, on account of the presence of the iris, al· though it does not protrude over the surface of the cornea, the scar cannot become strong enough to withstand even the normal intraocular pressure, and thus it begins gradually to bulge (*ectatic, cystoid scar*), and may, after a time, develop into a *traumatic staphyloma*. (See. Fig. 90). In rare cases the incarceration

Fig. 91.—Cystoid scar formed by atrophied iris tissue and corneal epithelium covering it.

may give rise to the formation of a *cyst* of the iris. The nearer the periphery of the iris the incarceration or the prolapse has taken place, the more serious are the consequences which may follow. The most serious of these are chronic plastic or purulent iritis and cyclitis, and finally sympathetic ophthalmia.

If the cut has penetrated both the cornea and the iris, the conditions will be much the same as in simple penetrating wounds of the cornea. There is, however, usually a large effusion of blood into the anterior chamber, which may interfere for a time with a careful examination.

If the cut is still deeper and penetrates the capsule of the crystalline lens, the situation is further complicated by the formation of cataract. If the wound in the lens-capsule is large the pressure from the rapidly swelling lens-substance, giving rise to acute glaucomatous symptoms, may very soon force us to attempt its extraction. In some rare cases the iris, without apparently being cut, may, by the injuring material, be driven into the crystalline lens, and may remain in this position, held tight by the lips of the wound in the lens-capsule. The ensuing dimness of the lens-substance may then remain confined to the immediate neighborhood of such a traumatic

posterior synechia.   In some cases the iris, or the iris together with the crystalline lens, are torn out of the eyeball
altogether by the instrument inflicting the injury (*traumatic
irideremia*).

Traumatic cataract may be removed by extraction, or in
young individuals it may be brought to absorption by means
of discission.   In aseptic injuries the cataract remains often
uncomplicated save for the small scar in the cornea.   When
the injury was attended with septic infection serious complications usually develop before the formation of the cataract or
with it.   Iritis and iridocyclitis may render the condition of
the eye less favorable for future operation or by their sequelæ
may make it impossible to give more than a small portion of
useful vision.   If deeper complications ensue the eye becomes
very painful, blind and dangerous to its fellow, and will
have to be removed.

Injuries to the slerotic happen but rarely without a contem-
poraneous injury to the ciliary body, or to the choroid and
retina.   When such a wound has been rendered aseptic and is
not gaping, rest and closure of the eye may accomplish all
that is required; but if the wound gapes widely, and the vitreous body shows itself in the opening, it is best to sew it up
by stitching the wound-lips of the conjunctiva over it, or those
of the sclerotic.   Sometimes the vitreous body is prolapsed
and the cut or torn edges of the choroid and, perhaps, also of
the retina, protrude between the lips of the wound in the sclerotic.   In such cases it is best to trim the edges of the
wound carefully and to stitch the sclera or at least the conjunctiva above the scleral wound.   After such an injury, in
spite of what is done, the eyeball is very frequently ruined,
and shrinks, and it may be a source of danger to the fellow
eye.

If a septic wound of the sclerotic lies in the ciliary region,
and the ciliary body is also wounded, septic cyclitis, in spite
of our efforts, is almost sure to follow.   An eyeball so injured
will shrink and is especially apt to cause sympathetic inflammation of the other eye.   In some cases the injury may involve almost all parts of the eyeball, and the eye will "run
out," or a chronic inflammatory process will lead to shrinking.

Injuries to the optic nerve are not often seen. They are direct cuts or tearing by bullets or other foreign substances penetrating into the orbit. They lead to partial or total atrophy of the nerve. When atrophy of the optic nerve follows an injury to the head, as from a heavy fall, etc., the atrophy is usually due to fracture of the walls of the canalis opticus.

§115. If the injuring foreign body, especially when septic, perforates the cornea or sclerotic and remains lodged within the eyeball, the injury is a particularly grave one, as stated.

The size of such a foreign body may vary considerably. Large pieces of metal, glass or wood will simply destroy the eyeball by the immediate injury they inflict while entering it. Small foreign bodies may act destructively in a variety of ways. If a foreign body is embedded in the cornea, it is easily removed with a scoop or needle. Care must, however, be taken not to push it through the cornea into the anterior chamber during the attempt at removal. If a small foreign body has entered the anterior chamber, it will usually remain entangled in the iris, or be embedded in the crystalline lens. If it remains in the iris, and cannot be removed in any other way, an iridectomy of the part which contains the foreign body ought to be made. If it remains in the crystalline lens, it usually causes simply the formation of a cataract, and it may be removed together with the lens-substance at a later period.

Sometimes a foreign body, after having struck the iris, will fall into the angle between the iris and cornea, and its removal from such a position is very troublesome, especially if it is not iron or steel and very small. In such a case it is well to move the foreign body with a needle into a position on the iris nearer to the pupillary edge, and then to remove it, with a portion of the iris, by iridectomy.

Particles of iron are best removed by means of a permanent magnet (*Gruening's*) or better by means of an electromagnet (*Hubbel's*).

If the removal of a foreign body in the iris or anterior chamber can be accomplished soon after it has entered the eyeball, the danger is generally averted. We should, therefore, remove such a foreign body as soon as its presence is known. In a few instances the presence of an aseptic foreign body in the anterior chamber has been borne well for a prolonged period, but such cases are very rare, and are altogether exceptional.

Small foreign bodies which have entered the vitreous body may be detected with the ophthalmoscope, and sometimes can be removed by its guidance. If the foreign body is a small piece of iron or steel, we may succeed in removing it by the aid of a magnet. If it is of a non-magnetic substance, a grooved hook (*Knapp's*) or smooth forceps may be used. For these purposes the sclerotic must be cut in a meridional direction, as near as possible, to the place where the foreign body is situated. If not removed, septic foreign bodies in the vitreous body give rise to suppurating panophthalmitis or, perhaps, to a lower type of inflammation, which may as much endanger the fellow-eye. In a few cases an aseptic foreign body, or one whose chemical decomposition did not later on act as an irritant, has been observed to remain in the vitreous body for years, doing no harm. Even when an eye is already inflamed from the presence of a septic foreign body in the vitreous, the removal of the offending body may sometimes arrest the progress of the septic inflammation.

Foreign bodies, which have become lodged in the ciliary body, almost certainly destroy the eyeball, and are also most frequently the source of the destruction of its fellow.

From the foregoing statements it is easy to understand that a penetrating injury to the eye, even when not complicated with infection is, in most cases, a very serious affair. When complicated by septic infection the gravity is still greater, as such an eye is in most cases lost and may destroy its fellow by sympathetic ophthalmia. (See Chapter XIX). The physician should, therefore, be extremely guarded with regard to the prognosis in all such cases, since even an apparently slight injury may turn out to have been a most serious one. We must be especially careful in examining an injured eyeball, and in doing so, as much as possible, avoid pressure upon it. If there is blood in the anterior chamber preventing further examination, the injury is probably a grave one. All that can be done, then, is to wait for the absorption of the blood, and meanwhile to prevent, as far as possible, the development of inflammatory symptoms by antiseptics, by enjoining strict rest in a dark room, and by instilling sulphate of atropia. If after a few days no inflammation has taken place and the blood is absorbed, so that the deeper parts can be examined, the eyeball as such may probably be considered as safe. If inflammation takes place, in spite of our precautions, the eyeball will usually be lost by suppuration.

Injured eyeballs, even if not altogether lost, especially when septic, are very likely to remain irritable or to gradually develop low forms of chronic inflammation, which may lead to to their destruction later on, and finally to the destruction of the fellow eye through sympathetic disease. Such eyeballs must, therefore, be constantly and closely watched so as to detect the danger in time to arrest or avert it.

It is plain that the subject of injuries to the eyeball is one of the greatest importance, and especially for the general practitioner, since he generally sees these cases first, and his actions and counsels generally determine the issue. Unless the injury is an absolutely superficial one, he should not undertake to give more than a doubtful prognosis. He should at once apply antiseptic measures, as stated above, and should prepare the patient for what may prove to be the only resource, not only to save him great suffering from the injured eyeball, but also to avert imminent danger of total blindness from the loss of its fellow, namely: The removal of the injured eyeball by *enucleation*. The patient will then probably more readily yield to the dire necessity, or if he does not, and finally becomes blind, the physician has at least done his duty.

The question, whether an injured eyeball is to be removed or not, depends not only on the nature and extent of the injury, but also on the fact whether a foreign body remains within the eyeball or not. Of this latter point the patient has usually no means of forming a correct judgment, and the physician should not rely on his statements, unless he can satisfy himself that the instrument which has caused the injury cannot possibly have left a particle within the eyeball. If there is no doubt remaining as to the necessity of removing the injured eye, the sooner it is done the better for the patient. The operation affords immediate relief from the excruciating and continuous pain, and the quiet and speedy healing, which is the rule after enucleation, will enable the patient to resume his work after a very short period.

When in any given case there is well-founded doubt, although still a probability that enucleation will become necessary later on, our action should depend largely on the patient's position in life, and on the possibility of watching the eyes carefully, and doing what is necessary at the first indication.

Instead of enucleation, evisceration, scraping of all the contents out of the scleral shell, has been introduced, as giving a better stump for the wearing of an artificial eye. The operation is followed by a slow and painful healing process. It is, moreover, less certain that all septic infective material may be removed by means of it, than by enucleation. To improve the stump still more, an artificial vitreous of glass (*Mules*) may be inserted into the scleral cavity and be allowed to become encapsulated. When there is only a question of good looks, this operation may be in place.

Section or removal of a piece of the the optic nerve and the ciliary nerves around its entrance (optico-ciliary neurotomy or neurectomy) have also their advocates. In view of a possible sympathetic ophthalmia they are not as reliable as enucleation.

From the great frequency and destructive nature of injuries of the eyeball incident to certain dangerous trades, the use of protective glasses cannot be too strongly urged upon the workmen, whose occupation exposes them daily to such perils. Such protective glasses are best made of mica, and in Europe, where they are extensively used, they have been the means of saving many a workman from blindness and many a family from destitution. It is greatly to be deplored that American workmen cannot be persuaded to use them.

§117. Injuries to the eyeball by blunt forces may cause rupture of the sphincter edge of the iris, iridodialysis, isolated rupture of the choroid and probably, also, of the ciliary body. These injuries may be accompanied by considerable hæmorrhage. If the hæmorrhage concerns the anterior chamber only, it is called *hyphæma*, if all the cavities of the eye are filled with blood it is called *hæmophthalmus*. Blunt force may furthermore cause the iris to be tilted backwards and remain in this position when it will look as if part of the iris had disappeared. (See Fig. 91). This is particularly observed when

FIG. 91.—Tilting backwards of the iris after contusion of the eyeball.

the force has struck the cornea directly from in front, as by a cork flying from a soda-water or champagne bottle. Sometimes a partial paralysis of the sphincter, probably due to the unbloody tearing of muscular fibres will render the pupil oval.

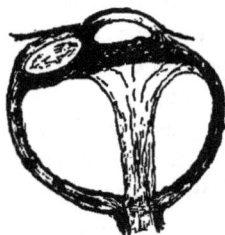

FIG. 92.—Dislocation of the crystalline lens under the conjunctiva by an injury which ruptured the sclerotic.

Subluxation and dislocation of the crystalline lens are often the result of blunt injuries. In rare cases the cornea or sclerotic is ruptured. In these cases the lens may at the same time be forced out of the eyeball, and it sometimes remains lying under the conjunctiva. (See Fig. 92).

§118. *Sympathetic ophthalmia* is the collective name given to all affections which are brought about in an eye by certain diseased conditions in its fellow, when these, and these alone, are the cause of the affection in the second eye.

If, for instance, a patient suffers from idiopathic iritis in one eye, and his other eye is attacked in the same way soon after, we do not call this a sympathetic iritis, because the second eye becomes affected through the same constitutional diathesis which has led to the iritis in the first eye.

By far the greater number of cases of sympathetic ophthalmia, if not all, are due to inflammatory processes induced in the first affected eye by an injury, with or without the continuing presence of a foreign body within the eyeball. Experience has shown that chronic cyclitis is especially apt to be developed in such an eye, causing often a similar sympathetic trouble in the other eye. From this fact it has been thought that a direct transmission of the inflammatory process takes place along the ciliary nerves. This theory is open to certain serious objections, and we have, moreover, other and more direct channels for the transmission of an inflammation from one eye to the other in the optic nerve with its sheaths and other lymph-channels. Pathological anatomy and experiments, as well as clinical observations, point decidedly to these channels as being the most important ones in the transmission of the disease.

Since the influence of septic bacteria upon the causation and transmission of disease has become known and carefully studied, the character of true sympathetic ophthalmia (not sympathetic irritation) as a transmitted septic infection has all but been established [*Leber, Deutschmann*] (*ophthalmia migratoria*). Although a number of observers have been unable to find the bacteria in eyes which had produced sympathetic ophthalmia (among them myself), this is no proof that *Deutsch-*

mann's views are incorrect. Whether the channels by which the migration of the septic bacteria takes place from one eye to the other are to be found in the optic nerve and its pia mater sheath alone (*Knies*), or also in lymph-channels which leave the optic nerve with the central bloodvessels and go through the orbit into the cranial cavity (*Gifford*), is still a mooted question.

The time at which sympathetic affection most frequently occurs is in from 4 to 6 weeks after the injury has been inflicted on the fellow-eye. In a large number of cases, however, such an injury may have preceded the occurrence of the sympathetic affection by many months, and even by years. In rare cases a few days only seem to have intervened between the affection of the two eyes.

As an eye once affected by sympathetic ophthalmia is, as a rule, ruined, and the patient is thus in most cases rendered utterly and hopelessly blind, this subject is one of the most important in ophthalmic practice, and the physician cannot be too deeply impressed with its importance. In most cases the duty will devolve upon him to forestall the fearful results, and to tell the patient what probably will be his only safeguard against utter blindness. If the physician does not recognize his duty but through lack of judgment, or for any other reason encourages the patient to reject the one effective remedy, the enucleation of an injured eyeball, or of an eyeball which, for other reasons, is likely to produce sympathetic inflammation, the blame will rightfully fall on his shoulders.

The eyes, which are most apt to give rise to sympathetic troubles are, as just stated, especially injured eyes, and among these again, especially eyes in which the injury has been in the ciliary region, or in which there has been a prolapse of the iris, or the ciliary body, or of the choroid and retina, and those eyeballs within which a foreign body has become lodged. Ectatic corneal scars, staphyloma in all its forms, plastic iridocyclitis, and iridochoroiditis, and anterior phthisis of the eyeball may also cause sympathetic trouble. Furthermore, operations on the eye, and especially cataract extractions, when followed by septic inflammation, may be the source of a sympathetic inflammation.

The primarily affected eyeball need not be absolutely destroyed to become a source of danger; it may even be a comparatively useful organ, and yet be so affected as possibly to cause a sympathetic affection in the other eye. In most cases, however, vision in the first affected eye is reduced to the mere perception of light, or is even altogether abolished before such an eye becomes dangerous to its fellow.

§119.   Sympathetic ophthalmia may appear clinically in different forms, which it is especially necessary to recognize in the initial stage.

The lightest form of sympathetic ophthalmia, and usually the forerunner of the more serious forms in which organic changes take place, is that which is conventionally termed *sympathetic irritation.*

An eye suffering from sympathetic irritation shows no organic changes; however, it can not bear the light well. It tires easily, especially in reading or similar occupations; moving the books farther off may give momentary relief (weakened accommodation). Soon even the slightest application of the eye to any work causes lachrymation and redness, and the attendant pain in the surrounding regions makes work utterly impossible. Sight may be at times slightly obscured, or the patient may see shining spots and flashes of light (*photopsia*).

In this stage of the disease, in which no anatomical lesions have apparently as yet taken place in the tissues of the secondarily affected eyeball, the enucleation of the eye which is the cause of the trouble, will generally be followed by a speedy recovery. We should, therefore, be very careful to instruct a patient, who is the unlucky possessor of an injured eye, or an eye that may at some time cause sympathetic inflammation, that such symptoms as have been enumerated, however trivial he may consider them, must not be overlooked, but must be promptly reported. He must, in fact, be made so thoroughly aware of the danger to his other eye that he will be startled at even the slightest unusual symptom in his good eye. All the symptoms which together we call sympathetic irritation are to be explained as caused by reflex-neurosis and are transmitted by the sensory, motor and sympathetic fibres by way of the ciliary nerves.

§120. In some cases these functional symptoms of sympathetic irritation have already become complicated by organic changes, and *sympathetic neuritis* or *neuro-retinitis* has actually set in when we first examine the eye. Even, then, if no further changes have taken place, *enucleation* of the first affected eyeball may bring about a perfect cure. Whether such a neuritis is a primary affection or whether it is due to a beginning inflammation of the uveal tract is not certain.

§121. More frequently, however, we observe that the patient is suffering from *sympathetic iritis*, which may be either of a serous or a plastic type. Serous iritis seems to give a comparatively good prognosis, but plastic iritis, which very soon develops into plastic *irido-cyclitis*, as a rule, leaves nothing to be hoped for. In a few reported cases in which enucleation has been performed as soon as the first symptoms of iritis have been detected, a cure has even then been effected, but in most cases it is useless, and it may even be injurious to enucleate at this period.

Gradually the sympathetic iridocyclitis develops into an *irido-choroiditis*, and the contraction of the plastic membranes leads to shrinkage of the whole eyeball with detachment of the retina and softening of the eye. Every chance of help is gone in this stage.

The process of sympathetic ophthalmia is, as a rule, very gradual; there may be times of apparent freedom from inflammation, or, at best, partial remission of the inflammation; but soon a new exacerbation will take place, and the destructive process goes on. In a few cases the disease stops before the eye is utterly ruined, and then a judicious operation may ultimately give some sight. But no operation on an eye, made useless by sympathetic ophthalmia, should under any circumstances be attempted, until all signs of inflammation or even of irritability are gone, or better yet, have been gone for some time. If, in such a case, the perception of light and the projection are good, and the intraocular tension is but slightly, or not at all reduced, an operation (usually iridectomy or iridotomy, or one of these combined with extraction of the frequently cataractous crystalline lens), may be undertaken with the hope of restoring some degree of vision. Any attempt at operation at an earlier period will be punished by a new exacerbation of the disease, or at best will prove useless, as even large openings made in the iris and the pathological newformations will be in a very short time closed again by inflammatory products.

§122. While the inflammation is in progress subconjunctival injections of mercury, mercurial inunctions, or mercury given internally, may prove of value; pilocarpine may perhaps, also, be of service. Untiring efforts are sometimes even at such a period crowned with a partial success.

Obstinate keratitis or scleritis has in some cases been caused by the presence of an injured or shrunken eyeball, and has been cured only after the enucleation of the offending organ, and is, therefore, described as *sympathetic keratitis* or *scleritis*.

It has been stated above that it has been recommended to substitute for enucleation the operation of division of the ciliary nerves and the optic nerve close to the posterior surface of the eyeball (*optico-ciliary neurotomy*), or the removal of a piece of these nerves (*optico-ciliary neurectomy*); also *evisceration* of the eyeball. (See Chapter XVIII). Enucleation, however, still remains the only really trustworthy remedy, and when it is performed in time and before any sign of sympathetic ophthalmia has appeared, the only safe prophylactic measure.

§123. The wearing of an artificial eye will, in a great measure, do away with the disfigurement caused by the enucleation as it not only appears very much like a living eye, but can even be moved to a certain degree in all directions. This is due to the fact that the external muscles of the enucleated eyeball grow together with the orbital tissue and the conjunctiva. The shell of the artificial eye, resting on the orbital fat covered with conjunctiva, is thus moved whenever one of these muscles contracts.

The wearing of an artificial eye is, however, a source of expense and often of annoyance. Discharges of the conjunctiva will accumulate behind the artificial eye and dry on its surface, and in order to clean it and wash out the conjunctival sack, it must frequently be removed. In cold weather artificial eyes are apt to break within the orbit. The expense and annoyance, therefore, make it desirable in certain cases to close the palpebral fissure after removing the cilia-bearing edges and, perhaps, also the tarsal tissue and conjunctival sack.

CHAPTER XX—ERRORS OF REFRACTION AND ACCOMMODATION.

§124. Every eye, which is so constructed, that, while it is perfectly *at rest, parallel rays* entering through its cornea *are united* in a point (*focus*) *on its retina*, is called an *emmetropic eye*. (See Fig. 93).

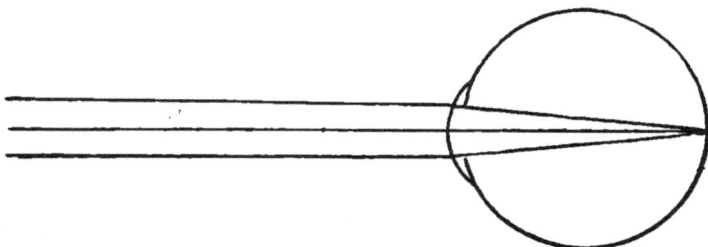

FIG. 93.—Emmetropic eye. Parallel rays passing through the refractive media of the emmetropic eye, when it is at rest, are focussed on the retina.

By parallel rays we mean, in practice, such rays as reach the eye from any distant object, and for most purposes we may, without material error, consider rays as parallel when the object from which they emanate is at any distance greater than twenty feet from the eye. The emmetropic eye sees, therefore, any distant object towards which it is directed distinctly and without effort.

Every eye which is not so constructed that, when in a state of rest, parallel rays are focussed on its retina, is called an *ametropic eye.*

An eye in which the retina *lies in front* of the focus for parallel rays, and which therefore cannot, in a state of rest, see even distant objects distinctly, is called a *hypermetropic eye* (over-sighted eye). (See Fig. 94).

Every eye whose retina *lies behind the focus* for such parallel rays, and which therefore sees near objects distinctly, is called 'a *myopic eye* (*near-sighted eye*). (See Fig. 95).'

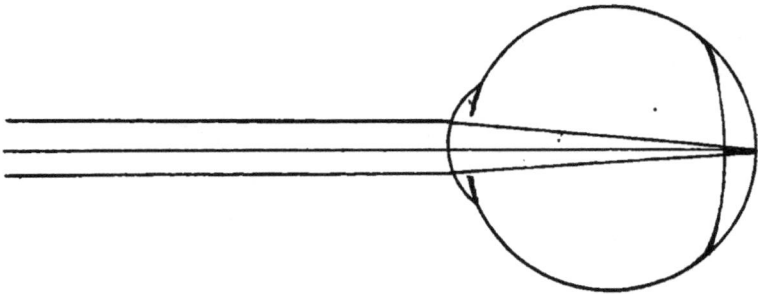

FIG. 94.—The hypermetropic eye is shorter than the emmetropic one and parallel rays are focussed in consequence behind the retina, when the eye is at rest.

In other words, *when perfectly at rest, the emmetropic eye is focussed for parallel rays, the hypermetropic eye is focussed for convergent rays, and the myopic eye is focussed for divergent rays.*

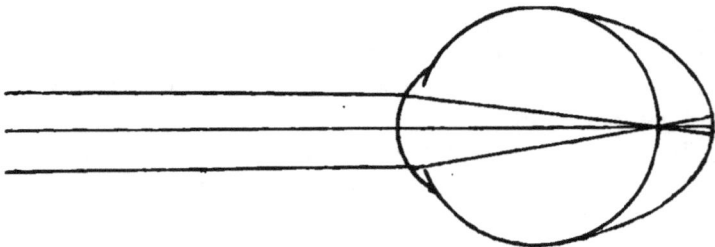

FIG. 95.—The myopic eye is an elongated eye. Parallel rays passing through its refractive media are focussed before having reached the retina.

The point for which an eye is focussed, when in a state of rest, is called the *far-point* of that eye. In the case of the emmetropic and hypermetropic eye the far-point is at an indefite distance; in the myopic eye the far-point is at a finite, and often a very short distance from the eye.

For the determination of the acuteness of vision in any eye (see Chapter II), we make use of test-types constructed after a certain principle. It has been found that an object, in order to be distinctly perceived by the human eye, must be seen under a visual angle of at least 1 minute. The letters, therefore, are arranged in such a manner that at a certain given distance each limb of a letter is seen under this visual angle, and the whole letter under an angle of 5 minutes. The letters are numbered to correspond with the distance in feet, at which each letter should be seen under this angle by the normal eye. Thus the letters which should be seen at 20 feet are marked

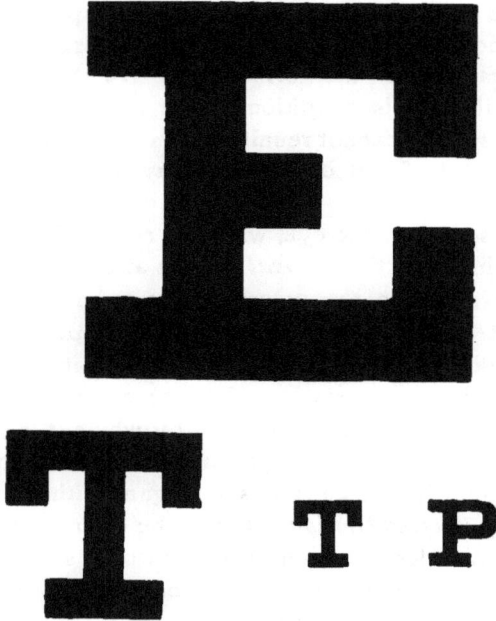

FIG. 96.—Snellen's test-types, seen by an eye with normal acuity of vision at two hundred, one hundred, and twenty feet respectively.

XX; at 70 feet LXX, and so on. (See Fig. 96). If at 20 feet distance an eye can read, for instance, only the letters which a normal eye ought to read at 70 feet, we express the visual acuteness (V) of that eye by the fraction $^{20}/_{LXX}$. (See Chapter II).

The same test-types are used for the determination of the refractive condition of the eyes. If an otherwise healthy eye

can see the letters, which are seen by the normal eye distinctly
at 20 feet at that distance, it is emmetropic, or it may, as we
shall see later on, be moderately hypermetropic. If such an
eye can read these letters as well or better, when a convex
lens is held before it, it is hypermetropic. If a concave lens
is required to bring vision up to $^{20}/_{xx}$, the eye is myopic.

Another form of ametropia is caused by an asymmetry
of curvature in the different meridians of the cornea, or
of the crystalline lens.   Such an eye sees everything
blurred and indistinct, and, although, perhaps, improved by
convex or concave lenses, it does not by their use alone come
up to the standard of the normal eye. We shall later on give
further details of this condition, which is called *astigmatism*,
because these eyes cannot reunite the rays which emanate from
any given point of the object upon any local point (*stigma*)
within the eye.

We have seen that the eye, when at rest, is focussed for its
far-point, which for the *normal eye* lies at an infinite distance,
but the eye has also the power of seeing small, near objects
with perfect distinction, or, in other words, it posseses a power
of adjustment by which it can focus upon its retina either
parallel or divergent rays.   This necessarily implies that
there is a faculty residing in the eye by which it is enabled, at
will to increase its refractive power to meet the requirements
of near vision.   This faculty lies in organs within the eyeball,
and is called *accommodation.* It may be expressed as equiva-
lent to a convex lens of such power as would suffice to render
the divergent rays coming from the near object parallel, as if
they came from a distant object.

The nearest point at which a small object can be seen distinctly by an eye, we call its *near-point.* The difference between the near and the far-point gives us the range of accommodation of an eye.

The accommodative power is, however, not the same throughout life. It diminishes with advancing age, and after the age
of 45 or 50 years it is reduced to a small fraction. In consequence of this loss of accommodative power, the near-point
gradually recedes farther and farther from the eye, and thus is
approximated more and more nearly to the far-point.

In every visual act not only must each eye be accommodated singly for the distance of the object, but the axes of the two eyes must be made to converge accurately upon the same point of the object in order to form identical images upon the maculæ luteæ of the two retinæ. Accommodation is, therefore, intimately associated with convergence, and whenever the one adjustment is called into activity, the other is performed at the same time, even when through some abnormal condition such an association of the two adjustments does not contribute to more perfect vision.

The organ by which the act of adjustment for near objects is performed, are the *ciliary muscle* and the *crystalline lens.* *Helmholtz* has explained the action in the following way: When the ciliary muscle, which forms a ring in which the crystalline lens is held by the suspensory ligament (*zonule of Zinn*) is contracted, this ligament becomes relaxed, and the crystalline lens, by its inherent elasticity, assumes a more nearly spherical shape, and its refractive power is correspondingly increased, as if a meniscus had been added to it.

When we observe an eye during the act of accommodation, we see that the pupil becomes smaller, and that the pupillary edge of the iris is slightly moved forwards. Accurate observation has shown that the increased convexity of the crystalline lens during accommodation is due mainly to a change in the form of its anterior surface.

§125. *Hypermetropia (over-sight or far-sight)* exists, as has been stated, when the eye in a state of rest is focussed for convergent rays, and parallel rays entering the eye are refracted towards a point lying behind its retina. In consequence of this condition the retina of such an eye receives only dispersion circles, and the images of distant objects, and still more of near objects, must be indistinct. This, as has been stated, can be remedied by *convex glasses,* and the convex glass which will allow a hypermetropic eye, when perfectly at rest, to unite parallel rays upon its retina gives us the degree of its hypermetropia.

In hypermetropia, if no glass is worn, the deficiency in refractive power is ordinarily, at least for distant vision, reme-

died by the exercise of the accommodation, and in low degrees
of hypermetropia, this may suffice for a time, even for near
objects. Low degrees of hypermetropia may thus remain unknown to the patient for years, or until his accommodative apparatus can no longer do its work effectively. This continuous strain of the accommodative apparatus causes, moreover,
a permanent contraction of the ciliary muscle, so that it can
no longer be perfectly relaxed at will.

In such a case, when testing the refraction, we may meet
with an eye with an apparently normal acuteness of vision,
although it is hypermetropic. That, what appears to be an
emmetropic eye is in reality a hypermetropic one, is recognized by the fact that a weak convex lens held before it, does not
diminish the acuteness of vision, and may even improve it.
The amount of hypermetropia which is thus disguised by the
help of accommodation is called *latent hypermetropia.* Even,
when the examination of an eye plainly reveals a certain degree of hypermetropia, the correction of which by means of a
convex lens, may give normal visual acuteness, an additional
qnantity of hypermetropia remains latent. The former is
called *manifest hypermetropia.* In order then to find out the
total hypermetropia we must get rid of all accommodative effort on the part of the patient. This is best done by *paralyzing
the accommodation* by means of the instillation of some drops
of a one per cent. solution of the sulphate of atropia, or one of
the more recent mydriatics. The *convex* lens which now suc-
ceeds in giving the patient normal acuteness of vision gives us
the exact degree of his total hypermetropia. The paralysis
of the accommodation by means of atropia renders the eye
unfit for any, or at least continued, near-work for from 10 to 12
days, and sometimes even longer. To obviate this the hydrobromate of homatropia has been introduced as a substitute,
as its action disappears in about thirty hours. After having
used this drug freely for several years in solution and in the
form of *Wood's* disks with cocaine added, I am satisfied that
in a large number of cases its paralyzing action certainly is
strong enough to replace that of atropine; but in a quite re-
spectable minority of the cases it is unable to reveal the total
amount of hypermetropia. Where time and other circum-

stances allow it, atropia should be used. When the patient cannot give the time, homatropine may take its place with reserve.

Some patients whose accommodative apparatus is strong enough to permanently overcome without discomfort a certain degree of hypermetropia, may never care to use correcting glasses, nor really feel the need of them until presbyopia develops. Others in whom there is perhaps a lower degree of strength of accommodative power than should even be expected in a normal eye, will require the correction by glasses of so small a degree of hypermetropia, that most individuals would never become conscious of its existence.

Theoretically, the glasses which correspond to the total degree of hypermetropia ought to be the best, or rather the most useful ones to the patient, since they will remove all undue strain from his accommodative apparatus. We find, however, generally that the old habit of accommodating more than necessary, comes back in some measure, when the effect of the paralyzing drug has passed off. The patients, therefore, often refuse these glasses at first, and are better satisfied with a number that lies between the degree of their manifest and of their total hypermetropia. Long continued paralysis of the accommodation may eventually do away with this habitual excessive strain of the accommodative apparatus, but it is generally better to begin by giving glasses of such strength as the patient can use with comfort, and to change them later on for stronger ones.

The question has even been raised whether in children it is really best to correct a moderate and even a medium degree of hypermetropia at all, and it has been pointed out that the accommodative strain during the years of development will help to bring about a natural cure of the hypermetropia by the stretching of the eyeball during its growth (*Carter*). In this manner a hypermetropic eye may become at some stage emmetropic. Such an occurrence is, however, just as apt to be followed by more stretching than is wished for and by such an increase in refraction that the eye becomes myopic. The gain under these circumstances would be a very questionable one.

Hypermetropic patients should, as a rule, wear their glasses always, except when their hypermetropia is of a very moderate grade, and their far-vision is comparatively good. In advanced years, when presbyopia is added to the hypermetropia, different glasses for distance and for reading must be used.

§126. The most frequent and characteristic symptom of hypermetropia is *accommodative asthenopia.* The patient may have perfect acuteness of vision, but, when doing near work, such as reading, writing, sewing, etc., his sight which at the beginning was good, becomes indistinct, the letters run into each other, and the eyes feel tired. By instinctively closing the eyes, and resting them for a moment, sight appears improved again; but soon this short rest fails to give relief. Later on, pain in the forehead, injection of the conjunctival bloodvessels and lachrymation are added to the former symptoms, and the work must be laid aside.

These symptoms are usually more pronounced when the near work is done by artificial light. The patients tell us sometimes that the symptoms have taken their origin from some severe illness, and in fact any weakening influence will suffice to bring them to light, by rendering the strain of the accommodative apparatus impossible.

Occasionally the patients are unable to read at all, and suffer from almost constant headache, and not infrequently we meet with a case in which the patient has been treated for years for some dark and hidden nervous trouble, and has been subjected to all sorts of needless deprivations and even operations, where the use of the proper glasses will suffice at once to remove all distressing symptoms.

Hypermetropic children often hold their books close to the eye, as if they were very short-sighted, and thus they strain their accommodation even more than would appear to be necessary for clear vision. This is probably due to the increase in size of the retinal image thus gained, and may grow with them into a confirmed habit. The fact, however, that such children can distinguish small distant objects clearly, such as birds, telegraph wires, etc., will at once reveal the fact that they are not short-sighted. In these cases a spasm of the accommodative apparatus takes place, in consequence of which the eye appears to be near-sighted and will apparently prefer correction by a concave glass, until the instillation of atropia and consequent paralysis of the ciliary muscle will show that the eye is really a hypermetropic one.

It has been stated above that the functions of accommodation and convergence are closely connected, and it is only after long practice that we can learn to accommodate without converging, or vice versa. The hypermetropic patient, in order to see small objects distinctly, must exert more accommodative effort than the emmetrope. He will, therefore, also instinctively converge his eyes more, and thus it may easily happen that he converges too much, so that *convergent strabismus* may be developed. In other words, the hypermetrope will, while using his accommodation in reading, converge his eyes more than is necessary for the distance of the object. He thus loses, of course, the benefit of binocular vision, and also sees the object doubled. As this diplopia is a source of confusion, he gets rid of it by turning one eye still further inwards, and works only with the other one. Thus the abnormal adduction is thrown entirely into one of the eyes, and the well-known picture of strabismus convergens (cross-eye) is developed.

Patients with a very high degree of hypermetropia do not gain very much in distinctness of vision by squinting; therefore, they, as a rule, do not fall into the habit. Patients with a very moderate degree of hypermetropia do not ordinarily need to sacrifice binocular vision to their fairly distinct perception of small objects. It is, therefore, chiefly in the medium degrees of hypermetropia, that strabismus convergens is observed.

From this causation of convergent strabismus it is evident, that there must be some time in the period of its development at which a proper correction by convex glasses and, perhaps, added to it the continued paralyzation of the accommodation and consequently of convergence, may bring about a cure. Such is, indeed, the case, when the patient is seen early enough, which, however, but rarely happens.

§127. *Myopia* (*short-sight or near-sightedness*) is, as stated above, that condition, in which the eye, when perfectly at rest, is focussed for divergent rays, and in which parallel rays entering through the cornea are united into a point before they reach the retina. This causes the retina to receive dispersion circles from far objects, and consequently such objects appear indistinct. On the other hand, near objects are seen clearly and without accommodative effort, but prolonged convergence is often made irksome by weak internal recti muscles.

Concave glasses enable myopic eyes to see distant objects distinct by rendering the parallel rays divergent and the slightly divergent rays more divergent, before they touch the cornea.

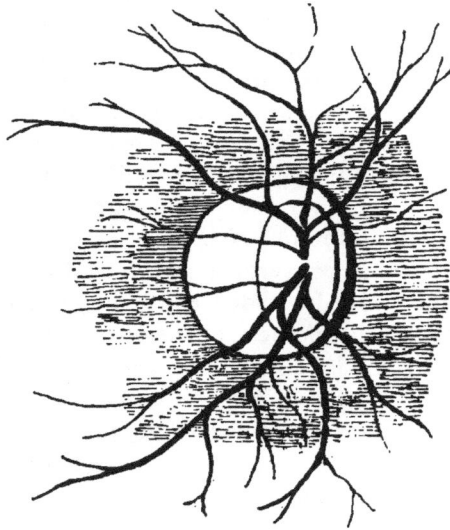

FIG. 97.—Ophthalmoscopic appearance of the optic papilla and crescent in a myopic
. eye.

In myopia the eyeball is too long in comparison with an emmetropic eye. This may be a congenital condition, or it may be acquired. The fact that among uncivilized nations myopia is almost unknown, and that it is pre-eminently an affection belonging to civilized life, shows us that its chief causes are to be sought in early and prolonged application to study, and especially under the unfavorable conditions of small print, badly ar-

ranged desks, which allow of or even actually lead the child
to assume a stooping posture, badly lighted, badly ventilated
and overheated school rooms, badly arranged or insufficient
artificial illumination, etc. Under such influences the child's
eyes, if there is any inherited tendency will certainly become
myopic.

§128. The elongation of the eyeball gives rise to certain
ophthalmoscopic changes in the background of a myopic eye.
The most constant symptom (although it may be observed
in eyes which are not myopic) is the appearance of a *crescent-*
shaped white figure, by which the disk is apparently enlarged
towards the macula lutea. (See Fig. 97). This crescent may
vary in its position and size, and the white figure may form a

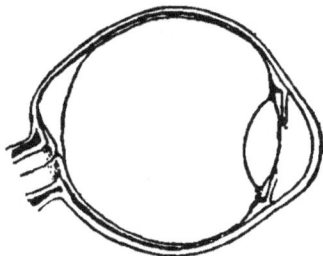

FIG. 98.—Shows how, by the formation of a posterior scleral staphyloma in progressive myopia, the vitreous body becomes detached from the retina. The
space so formed is filled by serum.

closed ring around the papilla. That this appearance is due
to a change in the choroid is evidenced by the fact that the
retinal bloodvessels pass over it undisturbed. When there is
not only an atrophic crescent, but a real congenital or acquired bulging of the posterior segment of the eyeball, we
speak of a *posterior scleral staphyloma.* By the gradual increase of such a staphyloma the vitreous body becomes detached and a space is formed between it and the retina, which
is filled by serum. (See Fig. 98). The staphyloma also leads
to a bend in the optic nerve at its entrance in such a manner
that the papilla seems to be dragged towards the macula lutea.

By this means the inter-vaginal space of the optic nerve near
the sclerotic becomes much wider than normal, and it penetrates sometimes far into the sclerotic.

§129. Myopia may remain comparatively or altogether stationary, or it may be progressive and become malignant. In
the latter case, aside from the staphyloma and atrophy due to
the stretching of the retina and choroid, symptoms of an inflammatory nature appear in the posterior part of the eyeball,
and characterize the process as a choroiditis. With every
new inflammatory attack new territory is invaded, and not infrequently the region of the macula lutea is thus rendered
blind. Floating opacities in the form of muscæ volitantes, or
larger fibrinous flocks, are never wanting and sub-retinal hæmorrhages may appear.

The pupil is usually larger in myopic than in emmetropic
eyes. To obviate the indistinctness of vision, due to the diffused light admitted through the large pupil, myopic patients
get into the habit of partially covering their pupils by squeezing the eyelids together or by actually flattening the cornea
by this pressure of the lids.

The elongation of myopic eyes causes them to appear full
and prominent.

§130. Although distant vision is very indistinct in myopia,
the sight for small near objects is excellent (at least where the
inflammatory changes have not gone too far), so that small
objects which an emmetropic eye can see only by the aid of
a weak magnifying glass, may sometimes be seen with ease by
a myopic eye. The amount of accommodative power used
by a myopic eye is very small, and in this, probably, lies the
reason why the converging (internal recti) muscles are also
weak and often insufficient for prolonged binocular vision. The
elongated shape of the myopic eyeball is furthermore a mechanical hindrance to convergence. This insufficiency of the internal recti muscles gives rise also to asthenopic symptoms due
to fatigue of the external muscular apparatus, and leading to
a loss of balance between the internal and the external recti,
so that at length one eye refuses the strain on the internal rec-

tus, and becomes passively everted by the unopposed action of the external rectus. To do away with the disagreeable double images which now appear and, in order not to give up the feeling of comfort caused by the relaxation of his converging muscles, the patient gradually allows the external rectus muscle more and more liberty of action, and finally a condition of permanent *divergent strabismus* is established.

Myopes, as they grow old, have a certain advantage over emmetropes and hypermetropes in their partial or total exemption from the necessity of using convex glasses for reading, but they do not, as is often erroneously assumed, become normal-sighted for the distance.

The concave glasses prescribed in myopia ought to be the weakest with which the normal acuteness of vision is obtained. The question whether a patient ought to wear his glasses constantly or only for distant vision, or whether he should use glasses of different strength for distant vision and for near work, depends mainly on the degree of his myopia, and upon the period of life at which the glasses are prescribed. The sooner accurately correcting glasses are worn, the better it is, as a rule, for the patient, and in as much as the corrected eye must use its accommodation for near work, it is usually best to encourage children to wear their correcting glasses constantly. In very high degrees of myopia and in advanced age a glass somewhat weaker than that which perfectly corrects the myopia, should be given.

Whenever a case of myopia shows decided signs of progressiveness or malignancy, prolonged rest from all use of the eyes should be enforced. To this may be advantageously added the instillation of atropia.

Synchisis of the vitreous body and detachment of the retina are apt to occur in myopic eyes, as has been already stated above.

§131. *Astigmatism,* asymmetry of curvature in the cornea (or crystalline lens), is the condition in which the different meridians of the refractive surfaces have unequal radii of curvature. To a slight degree this inequality exists even in the normal eye. When it is somewhat exaggerated, however, the sight becomes blurred, and details of objects appear more or less distorted for the reason that the unsymmetrical refractive surface can have no perfect focus for the rays which pass through it, but only a series of approximate foci lying along a line which has been given the name of focal interval. (See Fig. 99).

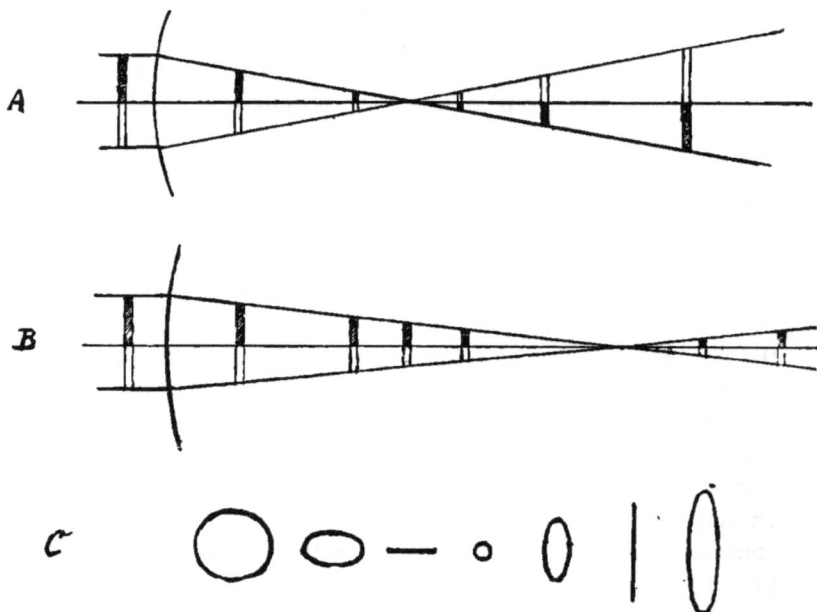

FIG. 99.—Refraction of parallel rays passing through the asymmetrically curved cornea in regular astigmatism. A. The refraction by the meridian of highest curvative. B. The refraction by the meridian of least curvature. C. Outline of a bundle of rays refracted by an asymmetrically curved cornea. The rays are united in two focal lines corresponding to the foci of the principal meridians. The space between them is called the focal interval of Sturm.

The name *regular astigmatism* is given to those cases in which the refracting surface is of a regular ovoid form, instead of being a segment of a sphere. In this form of astigmatism we recognize a meridian of greatest and one of least refraction, which two meridians lie usually at right angles to each other and are called the principal meridians. In astigmatism, according to the rule, the vertical meridian (or a meridian near the vertical one), is the more strongly curved, while the horizontal one (or a meridian near it) is the least curved one. When this condition is reversed, as we find in a number of cases, we speak of astigmatism against the rule. The astigmatic eye may be emmetropic in either of its principal meridians, in which case it will be either myopic or hypermetropic in the other, and we call this condition *simple myopic* or *hypermetropic astigmatism.* If the eye is myopic in both meridians it is called *compound myopic* or if hypermetropic in both meridians, *compound hypermetropic astigmatism,* and if it is myopic in one and hypermetropic in the other principal meridian, the condition is called *mixed astigmatism.* All forms of regular astigmatism may be corrected by means of a plano-cylindrical lens of the proper radius of curvature, so placed before the eye as to equalize its refractive power in the two principal meridians, and, if any ametropia remains, this may be corrected by grinding the proper spherical surface upon the other side of the same lens.

As a varying small amount of astigmatism exists in almost every eye, we may speak of normal astigmatism. It is due to this normal astigmatism that most of us do not see the stars as round bodies, but as stars.

While some patients will hardly be conscious of quite a marked degree of astigmatism and will even refuse to have it corrected by glasses, others suffer from a degree of asthenopia which seems out of all proportion to the small degree of astigmatism found on examination, and get relief only from its correction by glasses.

Regular astigmatism is nearly always congenital.

When the curvature of the refracting surface is irregular, we call the condition *irregular astigmatism.* This condition is mostly due to former inflammatory processes in the cornea.

Even in irregular astigmatism vision may sometimes be materially benefitted by spherico-cylindrical or cylindrical glasses.

§132. When the refractive power of the two eyes is not alike, the condition is called *anisometropia.* What glasses are to be used in such cases must depend on the special requirements of each case. It may be well to give a correcting glass

for each eye, if the difference is small; in other cases it is better to give glasses for the same focus for both eyes.   Even
when one eye is hypermetropic and the other myopic, a condi-
dition which might seem to be quite comfortable, some patients want a correction.   When the difference in the degree
of refraction is too great to give to each eye its correcting
glass, the glass which gives most comfort should be given.   In
correcting astigmatism each eye must be separately considered.

All forms of ametropia can be diagnosticated by the use of
the ophthalmoscope, when the patient's and the observer's accommodation are perfectly relaxed, and, if the observer be
himself not emmetropic, when his ametropia is corrected.

Other methods may be used to make the result of the examination more certain, and different apparatus have been devised by means of which we are enabled to make a perfectly
objective examination for astigmatism.   It does, however, not
lie within the scope of this book to do more than to refer to
their existence.   The two principal auxilliary methods are
skiascopy (*koroscopy, shadow-test*) and *keratoscopy*.   In the
former the direction and manner of the excursions which the
shadow of the iris will make across the lighted pupil when the
ophthalmoscope, through which it is viewed, is turned from
side to side, are made use of for measuring the refraction.   In
the second, the direction and degree of the distortion or displacement of an image reflected on the cornea are the means
of gaining a knowledge of the existence and degree of astigmatism.   The best instrument for the purpose of examining
for astigmatism is the ophthalmometer of *Javal & Schiötz.*

It cannot be expected that every physician shall be able to
make an exhaustive examination of the refractive condition of
a patient's eye.   Nevertheless he should be familiar with the
symptoms of ametropia, and especially of asthenopia, so that
he may advise the selection of glasses when needed, and,
moreover, see to it, that when prescribed, the glasses are worn
as directed by the oculist.   It is not only young ladies and
gentlemen who from vanity often refuse to wear glasses, even
though conscious that they are greatly benefitted by them.
Frequently the imperfectly educated parents will not allow
their children to wear glasses at an age when, perhaps, the

child's whole future may depend upon their use. The common prejudice against the wearing of glasses, and especially of convex glasses, by young persons, is not only unfounded, but it often leads to infinite harm. No one would refuse a patient with crippled legs the assistance of crutches, yet, to refuse the crippled eye the use of glasses is often a much greater wrong.

Furthermore, as asthenopsia of some kind may cause all sorts of otherwise unexplained nervous symptoms, among which *headache plays the chief rôle*, such symptoms should lead the physician to advise a thorough examination of the eyes before other methods of treatment are instituted. The cases in which such symptoms, due to asthenopia, are allowed to make a life miserable and to render the patient unfit for any application to near work, are still too commonly sent from one physician to another without relief, when a thorough examination of the eyes might readily lead to the recognition of the cause and its cure.

§133. When speaking of the *function of accommodation*, it was mentioned that with advancing years this faculty is gradually lost, and that this causes the near-point to recede more and more from the eye. Thus we find that while at 10 years of age the near point lies at about three inches from the eye, at 20 years it is about 4, at 30 years about 6, at 40 years about 9, at 45 years about 12, and at 50 years about 16 inches from the eye. From the fact that the range of binocular accommodation is somewhat less than that of monocular accommodation, the near-point for perfect binocular vision is even further from the eyes than these figures would indicate. The cause of this progressive loss of accommodative power lies in the physiological hardening of the crystalline lens, which renders it less and less capable of changing its form to meet the requirements of near vision (*presbyopia, old sight*).

As soon as the binocular near-point has receded beyond 12 inches from the eyes, reading, especially at night and for fine print, becomes uncomfortable. More light is required to see distinctly, and as the book must be held so far away as to allow the lamp to be easily placed between the book and the eyes, this remedy is usually at first resorted to. The comfort thus received is, moreover, due in part to the contraction of the pupil in bright light, and the consequent exclusion of dispersion circles.

§134. *Paralysis of the accommodation* causes the same symptoms as presbyopia. However, the pupil is in most cases at the same time perfectly dilated, while in other cases it may remain unchanged. Presbyopia always affects both eyes, paralysis of the accommodation may happen either in one or in both.

If paralysis of the accommodation affects a myopic eye, it will, like presbyopia, cause less inconvenience than when it occurs in an emmetropic eye. If it occurs in a hypermetropic eye, distant vision also becomes indistinct. Moreover, when in a case of paralysis of the accommodation, as is often the case, the sphincter pupillæ is also paralyzed, the diffuse light admitted into the eye through the dilated pupil, causes still greater confusion of sight than in a case in which the pupil is not increased in size.

Paralysis of the accommodation is a sign of some affection of the oculomotor nerve. It may be either an affection of the peripheral branches only, or the symptom of some central lesion. The most frequent cause of monocular paralysis of the accommodation is syphilis. Other causes of such a paralysis are diphtheria, tumors of the brain, poisoning with foul meat, fish, blood-sausage, and with certain drugs like belladonna, gelsemium, etc. Paralysis of the accommodation may also follow an injury to the eyeball.

The loss of accommodative power which occurs in connection with diphtheria, is generally incomplete, and is rather a paresis, than an actual paralysis. This paresis usually comes on several weeks after the diphtheritic process in the throat has run its course; it may be the only paresis following this disease, and it may appear conjoined with paresis of the muscles of the palate, etc. The inability to read in this affection is sometimes mistaken for obstinacy, and thus the child is liable to be misunderstood, and perhaps punished for his supposed fault; the physician should, therefore, always bear in mind that paralysis of the accommodation is a not infrequent sequel of diphtheria of the throat, and should warn the parents that the child's vision may possibly become affected, especially when other paralyses have developed.

In all such cases the child ought to be kept from school until perfectly well, although the use of a convex glass will enable him to read easily. Tonic treatment and rest will help to get him over the paresis, and, even if nothing is done in the way of medication, a few weeks' time will generally restore the accommodation. The instillation of mild miotic agents seems to shorten the time necessary for the recovery.

Spasm of the accommodative apparatus, leading to apparent myopia, is sometimes observed in hypermetropic eyes, and is usually very troublesome. If it happens in a myopic eye the degree of myopia is apparently increased. The treatment must be directed to the full relaxation of the accommodative apparatus, together with the correction of any existing ametropia.

CHAPTER XXI.—DISEASES OF THE EXTERNAL MUSCLES OF THE EYE.

§135. The eyeball can be moved upon its centre, in an infinite variety of directions. This is accomplished by means of the external muscles, the *rectus superior and inferior*, the *rectus internus and externus* and the *obliquus superior and inferior*. The four recti muscles, as stated in Chapter I (see Fig. 100), spring

FIG. 100.—(After Merkel). Shows the manner in which the external ocular muscles (except the obliquus inferior) take their origin from the bones around the optic foramen. R S. Rectus superior. R E. Rectus exterior. R I f. rectus inferior. R I. Rectus internus. O S. Obliquus superior. P S. levator palpebræ superioris.

from the apex of the orbit, around the optic foramen, and are inserted upon the sclerotic at different distances from the corneo-scleral margin. The superior oblique muscle also takes its origin at the apex of the orbit, and is inserted in the sclerotic, but only after its tendon has passed around the trochlea. The inferior oblique muscle springs from the inner surface of the orbit near its inferior and nasal margin, and then goes to the eyeball.

The action of the two oblique muscles is, of course, different from that of the recti, since their direction (See Fig. 101) from their punctum fixum (trochlea at the inner upper, and origin of the lower oblique at the inner lower orbital margin) is backwards and outwards. The distances from the corneo-

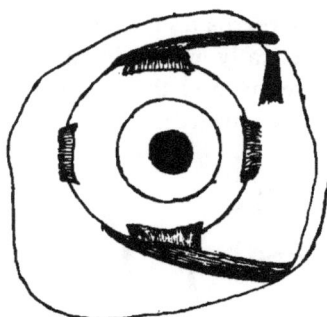

FIG. 101.—Manner in which the external ocular muscles are inserted on the sclerotic (right eye). The obliquus superior passes through a loop and thus its course is changed. The obliquus inferior springs from the lachrymal bone.

scleral margin, at which the several muscles are inserted in the sclerotic are, according to *Merkel*, the following ones:

Rectus superior, -	-	8.2 millimeters.
Rectus inferior, -	-	7.2 "
Rectus internus, -	-	6.5 "
Rectus externus, -	-	6.8 "
Obliquus superior, -	-	16.0 "
Obliquus inferior, -	-	18.3 "

These, of course, are average numbers. (See Fig. 102).

These muscles may act either singly or in various combinations. When acting singly the internal rectus turns the eyeball strictly horizontally inward, the external rectus in the same way turns it outward; the superior rectus which is inserted somewhat to the nasal side of the median plane of the eyeball, turns the eyeball upwards and slightly inwards, and rotates the upper end of its vertical meridian towards the nose; the superior oblique turns the eyeball downwards and outwards, and rotates the upper end of the vertical meridian of the cornea also towards the nose; the inferior rectus turns the eyeball downwards, and a little inwards, and rotates the upper end of its vertical meridian towards the temple; the inferior oblique turns the eyeball upwards and outwards, and rotates

the upper end of the vertical meridian of the cornea also towards the temple.

When the superior rectus and inferior oblique act together,
the eye is turned vertically upwards, and by the combined action of the inferior rectus and the superior oblique the eyeball
is turned vertically downwards.

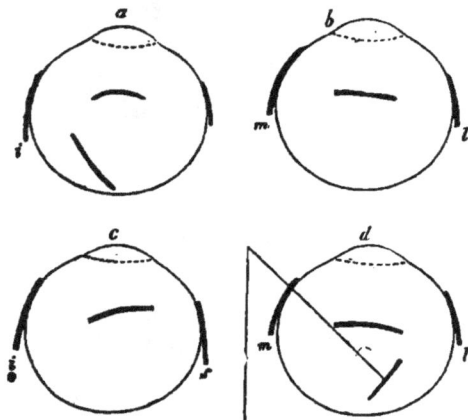

FIG. 102.—(After Merkel). Location of the insertions of the external ocular muscles.
a—i. Rectus inferior; s. Rectus superior. The other two lines refer to the
rectus externus and Obliquus inferior. b—l. Rectus externus; m. Rectus internus. The third line refers to the rectus inferior. c—i and s, as above.
The third line refers to the Rectus Internus. d—l and m, as above. The
other two lines refer to the Rectus superior and Obliquus superior and
show the angle at which the latter reaches the eyeball.

If the eyeball is turned upwards and outwards, the superior
rectus, the external rectus and the inferior oblique come into
play; if downwards and outwards, the inferior rectus, external
rectus and the superior oblique are in play, etc. Thus in all
movements, except those in the horizontal plane, at least two
and generally three muscles must act conjointly.

§136. When all the muscles of both eyes act in proper har-
mony, perfect binocular vision is the result, and two retinal
images are perceived by the brain as one object; but, if any
one muscle refuses to perform its part, or is paralyzed, the
two images will no longer fall on corresponding parts of the

two retinæ and double vision (*diplopia*) must follow. (See Fig. 103). The double or false image may be *homonymous*. It then

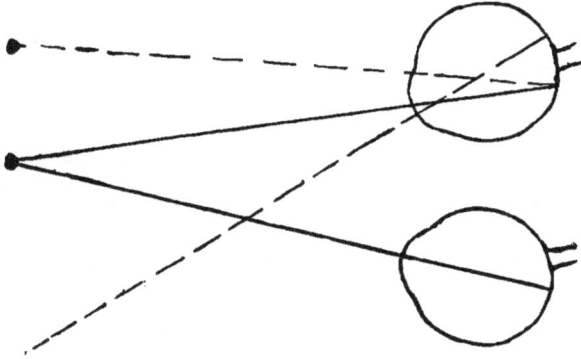

FIG. 103.—Homonymous diplopia in convergent strabismus.

appears on the side of the object looked at, corresponding to the eye in which the double image is perceived, or it may be *heteronymous* (*crossed image*), in which case it appears to be on the opposite side of the object. When one of the eyes is

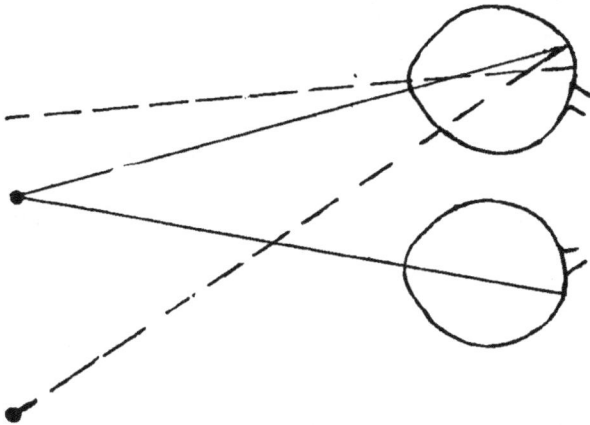

FIG. 104.—Heteronymous diplopia in divergent strabismus.

turned abnormally inwards towards the nose, the false image is homonymous; when one eye is turned abnormally outwards towards the temple, the false image is heteronymous. (See Fig. 104).

The oculomotor nerve directs the movements of all of the muscles of the eyeball except those of the superior oblique and the external rectus. These two muscles have each its own special nerve, namely: the trochlearis nerve for the superior oblique and the abducens nerve for the external rectus. The movements which the eyeball makes, when certain of its muscles act, being known to us, it is clear that the position of the false image will not only tell us what muscle or groupe of muscles refuse to act, but also which of the cerebral nerves may be affected. The diplopia is, therefore, often a symptom of great value in diagnosticating brain lesions, and may even help in locating them.

According to the degree of insufficiency of the muscle affected the false image will, in the positions of the eye which evoke it, appear nearer to or farther from the real image. The disagreeable consequences of seeing double are greatest when the distance between the two images is small; when this distance is very large, the false image is disregarded, or it may even not be seen at the same time with the real one.

In order to overcome diplopia, the patient will supplement the diminished action of the weakened muscle of the eyeball by turning his head in its direction. For instance, if the right eyeball cannot be abducted in consequence of paralysis of the external rectus, every object lying in this direction will appear double, as long as the patient gazes in this direction, but a turn of the head in the same direction will enable him to see single. From this position of the head, which after a while becomes habitual, we can often conclude in which muscle, or groupe of muscles, the affection lies.

§137. Diplopia may be brought about by any obstacle in the way of proper action of one or more muscles of the eyeball, as orbital tumors, orbital cellulitis or other causes, but it is most frequently produced by *paralysis* of one or more muscles, and this again is mostly due either to cerebral syphilis or to some other brain lesion. It has also been observed as the result of malarial poisoning. Paralysis of all external muscles of the eye, *ophthalmoplegia externa*, is seen in rare cases.

In cases of paralysis of one muscle of the eye we find after some time a secondary contraction of the antagonistic muscle just as we do in paralysis of other muscles of the body. The affection may thus bring about a form of strabismus which we call *paralytic strabismus*, to distinguish it from the typical or muscular strabismus. In paralytic strabismus the secondary deviation of the healthy eyeball, when covered, is greater than the deviation of the affected eye, while in typical muscular strabismus the primary and the secondary deviation are equal. (See Chapter II).

Paralysis of the external rectus is the most frequent cause of diplopia. If partial, the muscle may be able to move the eyeball somewhat beyond the medium line of the orbit, if total, it cannot move the eyeball at all, and convergent paralytic squint will soon follow. Diplopia must, of course, exist in all movements of the eyeball in which the external rectus ought to act. (The method of examining for double images has been detailed in Chapter II). The diplopia, characteristic of paralysis of the external rectus, appears, therefore, in the outer half of the binocular visual field. (See Fig. 105). The false

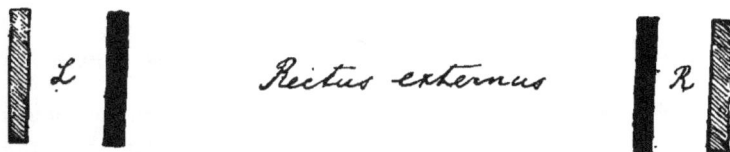

FIG. 105.—Position of the two images in paralysis of the rectus externus. The real image is black, the double image striated. R, right eye. L, left eye.

image stands on the temporal side of the real one. The distance between the two images increases, the farther the eyes are turned towards the side of the paralyzed muscle. In proportion, however, as the secondary contraction of the internal rectus becomes established, the diplopia extends over the whole field of vision.

Paralysis of the superior oblique causes the double images to appear when the patient looks below the horizontal line. The two images do not stand at an equal height, but the one belonging to the affected eyeball stands lower than the other and

converges towards it with its upper end. (See Fig. 106). The
vertical distance between the images increases when the eyes
are converged and look downwards, and the slanting direction
of the false image increases when the eye is abducted.

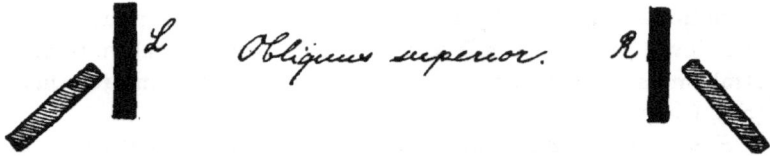

FIG. 106.—Position of the two images in paralysis of the obliquus superior.

If the *internal rectus* only is paralyzed, the double image
appears when the eyes are turned towards the healthy side,
and is on the nasal side of the true image. (See Fig. 107).
The more the eyes are turned towards the healthy side, the
farther the images are removed from each other.

FIG. 107.—Position of the two images in paralysis of the rectus externus.

Paralysis of the inferior rectus causes double vision below
the horizontal line. The false image stands lower than the
real one on its nasal side, and slants towards it with its upper

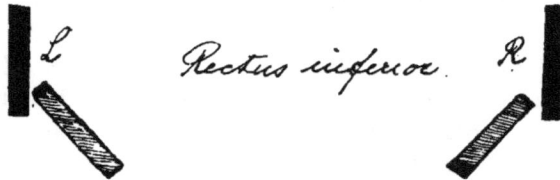

FIG. 108.—Position of the two images in paralysis of the rectus inferior.

end. (See Fig. 108). The images become farther removed
from each other in the vertical line, if the eyes are turned
downwards and abducted; the false image slants more during
adduction.

In *paralysis of the superior rectus* the diplopia is found above the horizontal line. (See Fig. 109). The false image stands above the real one on its nasal side, and slants towards it with its lower end. The vertical distance between the images increases when the eyes are turned upward and abducted; the false image appears more slanting when the eyes are converged.

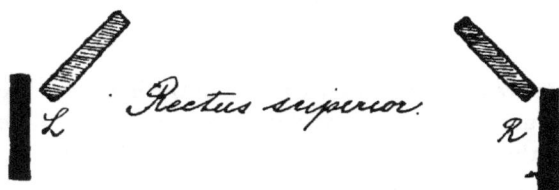

FIG. 109.—Position of the two images in paralysis of the rectus superior.

When the *inferior oblique* is paralyzed there is diplopia above the horizontal line. (See Fig. 110). The false image stands higher than the real one on its temporal side, and slants towards it with its lower end. The vertical distance between the two images is increased when looking up and converging. The false image slants more during abduction of the affected eye.

FIG. 110.—Position of the two images in paralysis of the obliquus inferior.

Paralysis of the sphincter pupillæ causes mydriasis (dilatation of the pupil) and is mostly combined with paralysis of the accommodative apparatus. (See Chapter XX).

§138. Paralysis of the oculomotor nerve may affect only one of its branches in the manner just detailed, but, particularly when it is of central origin, it affects several or all of its branches (levator palpebræ superioris, rectus superior, internus and inferior, obliquus inferior, sphincter pupillæ and ciliary muscle). When the sphincter pupillæ and ciliary muscle are affected alone we speak of *ophthalmoplegia interna*. In some cases, when the central lesion is progressive, a paralysis which originally affected but one branch of the oculomotor nerve may gradually affect the other branches one after another.

When the paralysis affects but one muscle it is usually either the external rectus or the superior oblique (trochlearis) since they have separate nerves. The prognosis is always a doubtful one.

§139. The treatment of paresis or paralysis of the muscles of the eye must take into consideration their probable cause. If syphilis is the general disease which has produced the paralysis, internal treatment by the proper drugs may be followed by success. To such internal treatment may be added the local application of the constant galvanic current. In same cases passive movements made with the eyeball in the direction of the paralyzed muscle, by means of a toothed forceps, yield good and prompt results; in other cases they prove of no value. When all chances of an improvement in the action of a paralyzed muscle have disappeared, a tenotomy of the opponent, or this combined with the advancement of the affected muscle may sometimes render the patient more comfortable. In other cases prismatic glasses may be used with comfort, when the affection has become stationary. (Sometimes they even help in the successful treatment of a paralysis). Whenever the patient's life is in jeopardy from the double vision, as in going down stairs or in walking on and crossing frequented streets and so on, or when he cannot get comfort in any other way, it is best to keep the affected eye covered.

§140. *Typical or muscular strabismus* almost always appears in early childhood, and is characterized by excessive muscular contraction. This prevents the patient from receiving the image of the object looked at upon the yellow spot in the two eyes. But, while in paralytic squint the eyeball cannot be moved at all in a certain direction, although there is binocular vision in other directions, the eyeballs move nearly equally well in all directions in the typical muscular squint, only they cannot both fix the same object at the same time, and there is, therefore, no true binocular vision. Double images, must, of course, exist in the beginning of muscular squint, but they are usually soon either disregarded and not noticed by the patient.

Muscular strabismus is practically either *convergent* or *divergent*, although upward and downward squint are seen in rare cases.

The causes to which the occurrence of the squint is usually ascribed by the parents are almost infinite in their variety, and it is not always judicious to try to correct their error. *Donders* first drew attention to the fact that an error of refraction exists in a very large proportion of the cases of strabismus. However, the number of cases of strabismus in which no appreciable error is found is not very small. A lack of equilibrium between the antagonistic muscles is, therefore, in these cases the only explanation. In another number of cases opacities of the cornea, or in the crystalline lens, or atrophic spots in the choroid and retina are the chief cause of the strabismus. In some cases an abnormal shape of the orbit, or an abnormal distance between the eyes may give rise to squint.

§141. Generally speaking *convergent squint* is most frequently found in combination with hypermetropia and hypermetropic astigmatism. From this fact it is clear, as pointed out above, that in such a case the timely use of correcting glasses may prevent the development of the strabismus. In practice, however, this is generally not the case, except when the ametropia is corrected, either before the strabismus has shown itself or at its very incipiency. Later on, when the strabismus has become established, the glasses will no longer cure it, or perhaps only after having been worn for a very long period. All other contrivances, such as opaque spectacles with a small central hole, or spectacles with one-half of each glass ground dim, etc., are of little or no value for the cure of strabismus.

In a great many cases of strabismus we find that one eye (the deviating one) is amblyopic, and the question is, whether this amblyopia is congenital, and has existed before the squint, or whether the fact that this eye is not used (at least consciously) brings about an amblyopia *"ex anopsia."* While the former is surely often the case, I do not think the latter should be considered improbable (*Schweigger*, *Noyes*), as we know of many analogies to it in other parts of the human body.   Moreover, not infrequently a crossed eye is observed to become decidedly less amblyopic after an operation for strabismus.

In the beginning strabismus is usually only periodical, and may occasionally remain so for months, or even years; sooner or later, however, it almost always becomes permanent.

Convergent strabismus oftenest makes its appearance within two or three years after birth, when the child begins to employ the eyes in looking intently at small, near objects, toys, pictures, letters, etc.   It is said to be sometimes congenital, but this is, to say the least, very rare.   When in convergent strabismus the squinting eye moves with the unaffected one and to an almost equal extent, we speak of *strabismus convergens concomitans*.

§142. *Divergent squint* is usually associated with and dependant on myopia and myopic astigmatism.

Before the divergent strabismus becomes pronounced, insufficiency of the internal recti is always observed.   In this condition the convergence of the eyes in near vision is imperfectly maintained, so that the visual axes become relatively divergent, although in distant vision they may be apparently or actually parallel.   Divergent squint, develops, as a rule, only slowly.   It is, moreover, apt to come on at a more advanced age than that at which convergent strabismus usually appears.

Insufficiency of the internal recti may also exist without ever developing into divergent strabismus, and in these cases the patients can, although with an undue effort, converge enough for binocular vision.   This causes athenopic symptoms (*muscular asthenopia*) to appear, which are often as distressing as the asthenopia which depends upon the undue exertion of the accommodative apparatus in hypermetropia.

Such insufficiency of action may be observed in all muscles of the eye, and asthenopic symptoms, may, therefore, be due to the insufficiency of other muscles than the internal recti.

Such a muscular asthenopia, as well as the accommodative

asthenopia (Chapter XX), may give rise to reflex nervous phe-
nomena, especially severe headache when near work is attempted. That this is the case cannot well be doubted, but in
recent years this influence has most certainly been over-rated
by some authors, so that but few functional nervous diseases
are in existence of which muscular asthenopia has not by some-
one been declared to be the cause.

§143. Permanent strabismus can be remedied by operation
only. This operation is *tenotomy* or division of the tendon of
the contracted muscle of the squinting eye, and when necessary also of the same muscle of the other eye. In other cases
it may be best to combine tenotomy on one eye with *advancement of the tendon* of the antagonist, or the advancement of
the antagonist of the contracted muscle alone may suffice. A
more modern procedure is the *advancement of Tenon's capsule*
(*von Wecker*) over the antagonist.

Whenever the strabismus is permanent and can no longer
be beneficially influenced by glasses, it is time to perform these
operations in order to obviate the amblyopia *ex anopsia*. The
age for operation, therefore, depends on the lack of probability of a change by other means. When it has to be done in
early childhood, it is well to confine the tenotomy to one eye
and to carefully avoid undue cutting of tissues. Cases in
which a well developed strabismus will disappear without surgical interference are extremely rare, although they occur
occasionally.

Insufficiency of any of the external muscles of the eye may
be improved by orthoptic treatment by means of prisms. If
no improvement can be gained in this manner the asthenopic
symptoms may be relieved by the constant wearing of prismatic glasses (or the decentration of spherical lenses when
such are worn). In some extreme cases no relief can be obtained by these means and an operation has to be performed.
This is best done by total tenotomy of the comparatively too
strong muscle. Partial tenotomies have been introduced of
late by *Stevens* and have found their admirers.

§144. *Nystagmus* is a continuous motion of the eyeballs

mostly in a horizontal but sometimes in a rotary direction (pendulum motion). It is generally observed when central vision is poor, as in cases of congenital amblyopia, retinitis pigmentosa, cataract, scars on the cornea, albinism, etc., or it may develop in advanced life as a result from brain disease. It is also observed to attack miners, and is then due to the cramped position in which these men have to work and not to the mode of illumination and poisonous gases as has been thought (*Snell*).

Treatment is useless, except in miner's nystagmus, in which form rest, fresh air and, perhaps, tonic treatment usually affect a cure.

CHAPTER XXII.—ON THE DIAGNOSTIC VALUE OF EYE-DISEASES IN INTRA-CRANIAL AFFECTIONS.

§145. It is not very long since it was thought that by the use of the ophthalmoscope an observer could look, so to speak, into the brain, or, in other words, that from the conditions of the optic nerve and retina we could with certainty infer what was going on in the brain. Thus cerebroscopy seemed about to be established as a new branch of medical science. We have learned since that the facts do not warrant such enthusiastic views, and we have now a comparatively clear knowledge of what we can expect from ophthalmoscopic and other eye-symptoms as aids to the diagnosis of brain affections. There remains no doubt that some eye-symptoms help us to diagnosticate certain intra-cranial lesions, and they are sometimes even of very great diagnostic value. In a few cases they may even enable us not only to diagnosticate an intra-cranial affection, but also to locate it.

Without going further into anatomical details we may here state, that the optic nerve and retina are really parts of the brain, from which they grow during fœtal life, and that their blood-vessels and their lymphatics are directly connected with those of the brain. Furthermore, we must keep in mind that the sheaths of the optic nerve are directly continuous with the meninges, and that the intervaginal spaces of the optic sheaths correspond to and are in direct communication with the spaces bounded by these membranes within the cranium.

It is evident from these anatomical facts that an increased or diminished supply of blood or lymphatic fluid in the brain and cranial cavity must, when no other affections exist, cause a like condition in the optic nerve and retina.

Although these facts are simply due to the mechanical conditions and cannot be doubted, their diagnostic value is small-

er than we should expect, because even in normal eyes the individual differences in the number, the situation and the form
of the bloodvessels as seen with the ophthalmoscope, are
such that we can hardly venture to diagnosticate a small degree of anæmia or hyperæmia, unless we have had occasion to
examine the eyes at a former period. Yet, it is in just these
cases of incipient hyperæmia and anæmia, that the diagnosis
might often be of the greatest value.

When, however, the anæmia or hyperæmia of the optic
nerve and retina have reached a high degree, we can easily
recognize them, and thus, in combination with the other symptoms present in the case before us, they may help to a diagnosis.

If, for instance, the general symptoms lead to the conclusion that there must be hyperæmia of the brain, a pronounced
hyperæmia of the optic nerve and retina will confirm this diagnosis, if all other causes for the hyperæmia of these parts
can be excluded.

Or, if we find in an otherwise healthy individual, pronounced
anæmia of the optic nerve and retina (in a case of injury to
the head, for instance), the diagnosis of anæmia of the brain
may safely be made.

In other cases the ophthalmoscopic diagnosis of hyperæmia
or anæmia of the optic nerve and retina, and consequently of
the brain may help us even in the diagnosis af a further affection. For instance, if we find in a case of pertussis (whoop-
ing-cough) that the optic nerve and retina are perfectly anæmic, we know that the brain is also anæmic, and this, in connection with the knowledge that the patient's system has been
greatly reduced by the disease, may help us to the conclusion
that the heart's action, especially, must be very weak. The latter is, furthermore, proved by the fact already mentioned, that
by paracentesis of the anterior chamber and the consequent
lowering of the intraocular tension, we can bring about a re-
filling of the bloodvessels of the optic nerve and retina.

§146. Although pronounced hyperæmia, or anæmia of the
optic nerve and retina must necessarily happen quite often, we
have much oftener occasion to observe and to utilize for diag-

nostic purposes certain more pronounced changes of the tissue, namely, œdema of the optic papilla and optic neuritis, for the reason that they are usually combined with more or less important disturbances of vision.

Any intra-cranial affection which causes an increase of the intra-cranial pressure must also cause an increase of pressure in the intervaginal spaces of the optic nerve. This increase of pressure will lead to a dropsical distension of the sheaths of the optic nerv near its entrance into the eyeball (See Fig. 111)

FIG. 111.—Distention of the intervaginal space of the optic nerve by intracranial fluid in a case of intracranial tumor.

with venous stasis and œdema of the optic papilla; soon these conditions cause inflammatory symptoms, first in the optic papilla, and then also in the retinal tissue. Whenever, therefore, we find neuro-retinitis in a case in which it is due to an increase of intracranial pressure, it must have been preceded by œdema of the optic papilla.

FIG. 112.—Distention of the intervaginal space by organized tissue, the result of fibrino-plastic exudation driven into the intervaginal space in a case of meningitis.

Neuritis optica may, however, furthermore, be caused by a fibrinous or fibrino-purulent inflammation of the sheaths of the optic nerve based on a similar form of meningitis. This may be due to the fact that an exudation resulting from meningitis may simply be forced into the intervaginal space of the optic nerve (See Fig. 112), or the inflammatory process may itself spread from the intracranial meninges to those of the optic nerve.

The result of these inflammations, as we have seen above, is in most cases the total atrophy of the optic nerve.

Œdema of both optic papillæ, or neuro-retinitis in both eyes, give us, therefore, generally a hint, that there is an increase of intracranial pressure. This pressure is, however, of little further diagnostic value (although this hint alone may under some circumstances be of very great importance), since intracranial tumors, hæmorrhages, abscesses, and encephalitic and meningitic process may all cause these symptoms at the ocular end of the optic nerve. Yet, we know that in by far the largest number of cases these symptoms are due to an intracranial tumor, and this fact, together with the general symptoms, as well as other functional troubles which may, perhaps, exist in the eye, will in most cases help to make the diagnosis sure.

The diagnostic value of the condition of the pupil is mainly confined to the fact that paralysis of the oculomotor nerve or irritation of the sympathetic nerve cause dilatation of the pupil. Irritation of the oculomotor nerve or paralysis of the sympathetic nerve cause contraction of the pupil.

§149. A subjective eye-symptom which has an important bearing on the localization of cerebral affections is called *hemianopsia (half-blindness)*. In this condition one-half the visual field of one or both eyes is wanting.

When examining the visual field of an eye we must bear in mind that the nasal half of the visual field corresponds to the temporal half of the retina, while the temporal half of the visual field is that of the nasal half of the retina. In the same manner what is upwards in the visual field is perceived by the lower parts of the retina, and what lies downwards in the visual field is seen with the upper half of the retina.

The diagnostic value of hemianopsia in cerebral lesions is particularly due to the *partial decussation* of the optic nerve fibres when forming the optic chiasma.

The course of the optic nerve fibres in man, as at present accepted, is the following:

There is an optic center in the occipital lobe of each half of the brain, situated in the cortex more particularly in the region of the cuneus and near the gyrus angularis. From here the radiating fibres (of *Gratiolet*) go forward in the internal capsule to the pulvinar of the optic thalamus, are joined by fibres from the corpora quadrigemina and geniculata, and with these form the tractus opticus of each half of the brain. When the two

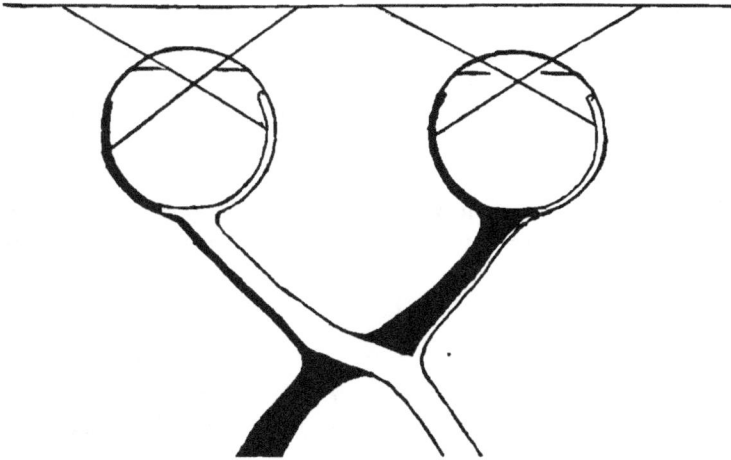

FIG. 113.—Shows the partial decussation of the fibres of the optic nerves while forming the optic chiasm. (The uncrossed bundles are too thin in this drawing). The lines projecting from the retina to the horizontal line refer to
the visual fields.

optic tracts reach each other in their forward course the optic chiasma is formed by the partial crossing over of the nerve fibres of the right hemisphere to the retina of the left eye, and of the fibres of the left hemisphere to the retina of the right eye. This partial crossing takes place in such a manner that the crossed bundle from the right tract (about two-thirds of the tract) reaches and supplies the right third of the retina of the left eye, and in consequence leads to the optic center all im-pressions perceived in the left portion (half, as it is called) of the visual field of the left eye. (See Fig. 113).

On the other hand, the uncrossed bundle of fibres from the
right optic tract goes to the right two-thirds of the retina of
the right eye, and conveys to the cortical center all impressions received in the left portion (half, as it is called) of the
visual field of the right eye.

Thus, the crossed bundle from the left optic tract supplies the left half of the right retina, and the uncrossed bundle the left half of the left retina and, of course, the opposite
parts of their visual fields.

From the foregoing it is apparent, that a lesion interfering
with the perception at the optic center itself or with the con-
duction of the optic fibres in one side of the brain, during their
course between the optic center and the chiasma, will cause
partial or total hemianopsia, that is, loss of a part of half or
the whole half of the visual field on the opposite side in both
eyes.

This form of hemianopsia is the most frequently observed
one and it is called bilateral *homonymous hemianopsia* (right or
left sided). (See Fig. 114).

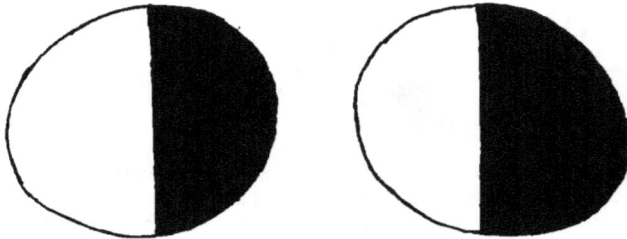

Fig. 114.—Homonymous (right-sided) hemianopsia. The dark parts are wanting in
the visual fields.

Such a lesion may be a tumor, an abscess, a hæmorrhage, a
trauma, an encephalitis or a meningitis, and if it does not involve any motor or sensory area, the case may simply present
the symptoms of hemianopsia. If the lesion lies where the
optic tract lies near any motor or sensory centers and involves
them at the same time, symptoms of motor or sensory paralysis or paresis or perhaps of spasms and hyperæsthesia may be
combined with the hemianopsia.

Should hemianopsia be combined with soul-blindness alone we might, perhaps, surmise correctly that the lesion must lie in the optical center in the occipital lobe where *Munk's* center is also located. If we had a case in which hemianopsia would be combined with paralysis of some of the external muscles of the eyeball, we should probably rightly conjecture a lesion at the base of the brain, etc. With regard to the diagnosis of the nature of the lesion in all these cases the symptom of hemianopsia is, of course, not calculated to give us much enlightenment.

Besides the bilateral homonymous hemianopsia, cases of bilateral *heteronymous hemianopsia* and *unilateral hemianopsia* have been observed.

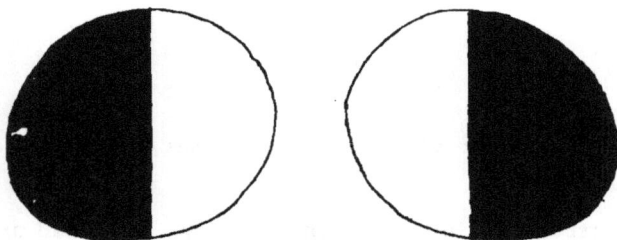

FIG. 115.—Heteronymous (in this case "temporal") hemianopsia. The dark. outer, halves are wanting in the visual fields.

Bilateral heteronymus hemianopsia is either temporal or nasal. The temporal form must be due to a lesion encroaching upon the anterior commissure of the chiasma, that is the crossed bundles of the optic nerve fibres. (See Fig 115). If the lesion only affects the crossed bundle of one side unilateral temporal hemianopsia must result.

Nasal bilateral hemianopsia can only be due to a lesion which involves both lateral commissures of the chiasma, that is, the uncrossed bundles of the optic nerve fibres.

CHAPTER XXIII.—DEVELOPMENT OF THE EYE AND CONGENITAL MALFORMATIONS.

§150. The time when the formation of the eye begins is not established for man. It seems, however, that it takes place between the third and fourth week.

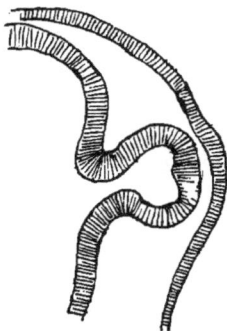

FIG. 116.—Primary ocular vesicle springing from the first cerebral vesicle on the right.

The first sign of the formation of the eye is a small protrusion of the lateral wall of the first cerebral vesicle. (See Fig. 116). This protrusion is called the *primary ocular vesicle*. The primary ocular vesicle is covered by mesoderm and ectoderm. The ectoderm at the height of the primary vesicle begins soon to grow thicker and convex towards the inner side. As it grows an indentation results at first in the primary vesicle and gradually as the ectoderm grows farther and farther inward, the *secondary ocular vesicle* is formed by the re-duplication of the primary vesicle. (See Fig. 117). During this period the ocular vesicle is gradually farther removed from the cerebral vesicle and the communication between the two, at first quite wide, grows narrower and longer. In this hollow

pedicle later on the optic nerve is formed. The swelling ectoderm by which the formation of the secondary ocular vesicle is
brought about is in the beginning open toward the surface.
Gradually this opening is closed and the sack so formed filled
with ectodermic elements, is later on the crystalline lens. It is
now covered by mesoderm and ectoderm. During this period
the ocular vesicle has an opening on the lower side in the
shape of a deep furrow, the *fœtal ocular fissure.* Through this

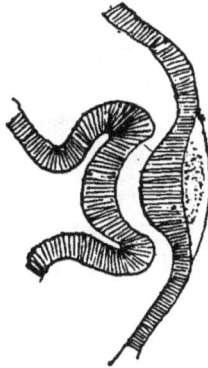

FIG. 117.—Secondary ocular vesicle. Beginning of the formation of the crystalline
lens.

fissure, which extends back into what is later on the optic
nerve, mesoderm enters between the lens and the ocular vesicle and forms the vitreous body, which is soon filled with
bloodvessels. Meanwhile the two layers of the ocular vesicle
have changed in such a manner that the external layer has
become thin and pigmented, while the inner layer has grown
quite thick. (See Fig. 118). The outer layer forms the pigmented epithelium, the inner layer the retina. Bloodvessels
appear in the mesoderm surrounding the whole ocular vesicle.
They form the choroid in the posterior parts, and in front of
the crystalline lens, the vascular pupillary membrane. Finally
sclerotic and cornea become differentiated. The ciliary body
and iris grow toward the axis of the eye from the part where
the external layer of the primary ocular vesicle is folded in-

ward to form the secondary vesicle.   The fœtal ocular fissure
closes in the eyes of mammals at the beginning of the second
month.

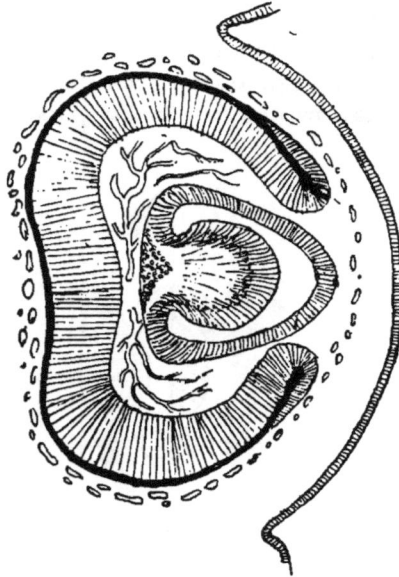

FIG. 118.—The outer portion of the ocular vesicle forms the pigment epithelium, the
inner one the retina.   The crystalline lens lies in front of it.   Between
the retina and crystalline lies the vascular vitreous body.   The blood-
vessells surrounding the ocular vesicle form the choroid and the membrana pupillaris.   (After J. Arnold).

The bloodvessels in the vitreous body come from the arteria
hyaloidea, a branch of the *central retinal artery*.   This larger
bloodvessel runs forward through the vitreous body (*canal of
Cloquet*) to the posterior surface of the crystalline lens (*membrana capsularis*).

These bloodvessels, as well as those in front of the crystall·
ine lens (*membrana pupillaris*), disappear in about the sixth
month or later.

§151.   The eyes and their adnexa are not unfrequently the
seat of *congenital malformations*.   These may affect only one

eye or both and may be due to faulty development, to arrest of development, or to diseases during fœtal life, or they may be caused by the persistence of parts which, though playing an important rôle during the development of the eye, have usually undergone retrogressive metamorphosis and and have disappeared before birth. The anomalies concerning the whole eyeball or several parts of it, due to arrest of development are dependent in the main on the too early or too late and insufficient closure of the fœtal fissure and date, therefore, back to an early period of fœtal existence. Anomalies concerning the iris, lens, or cornea alone, and the adnexa of the eye are usually formed at a later date of the development of the fœtal eye.

§152. The most frequent malformation which concerns the whole eyeball is that it is relatively too small (*microphthalmus*); this is often combined with other anomalies, as *microcornea*, coloboma of the iris or of the iris and choroid, cataract and dislocation of the crystalline lens. It mostly concerns both eyes. While vision in some cases is very good, in others it is but poor. In a great number of cases it has been seen combined with ptosis, probably due to the smallness of the eyeball which renders the action of the levator much less efficient.

§153. Sometimes one or both eyeballs are considerably too large, *megalophthalmus* (*buphthalmus, hydrophthalmus*). In this malformation the vitreous body is not gelatinous but fluid, and the intraocular pressure is, or was at one time, considerably above the normal. Sometimes the anterior chamber and cornea (*megalocornea, cornea globosa*) are more especially increased in size (*hydrophthalmus anterior*). The cornea in these cases may be perfectly transparent, or it shows partial dimness. The iris is usually discolored and atrophied, and trembles when the eye is moved. In the latter case the suspensory ligament is stretched, or has given way in some portion, and the crystalline lens is partially dislocated.

The cornea is sometimes the seat of congenital leucoma without any other malformation. Sometimes its shape is that of a cone (*keratoconus, conical cornea*). (See Fig. 48, Chapter

VIII). At the periphery of the cornea we sometimes meet with a small roundish tumor bearing hair, and consisting of the tissues of the skin (*dermoid*).

§154. Sometimes the iris is totally absent or nearly so (*aniridia, iridemia*). In other cases only a sector of the iris is wanting (*coloboma iridis*). (See Fig. 119). This sector is usually wanting in the lower half of the iris, the region of the fœtal fissure. The coloboma may reach further back into the ciliary body and choroid, retina and optic nerve. Such colobomata have also been found in the crystalline lens and the vitreous body. It is, furthermore, most likely that the posterior staphyloma (staphyloma of *Scarpa*) in myopic eyes has its origin also in an insufficient closure of the fœtal fissure. Its position can, however, only be explained by a rotation of the fœtal eyeball around its axis.

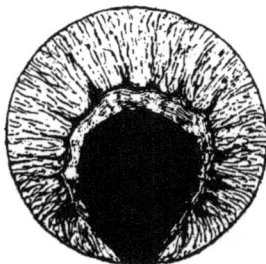

Fig. 119.—Congenital coloboma of the iris.

Sometimes the pupil is found to lie excentrically (*korectopia*). Remnants of the fœtal pupillary membrane are found quite frequently. They may be small threads hanging from the anterior surface of the iris into the pupillary area, or such threads may cross over the pupil from one side of the iris to the diametrically opposite one. (See Fig. 120). When these remnants are broader bands of tissue and anastomose with each other in the pupillary area they may divide it, so to speak, into several pupils (*polykoria*), and as many as five such pupils and even more have been observed in one eye.

§155. Another remnant of fœtal life, the *arteria hyaloidea*, is not rarely found. It appears as a string of abnormal tissue

FIG. 120.—Persistent remains of the fœtal pupillary membrane.

attached to the optic papilla and reaching into the vitreous body, sometimes as far forwards as the posterior surface of the

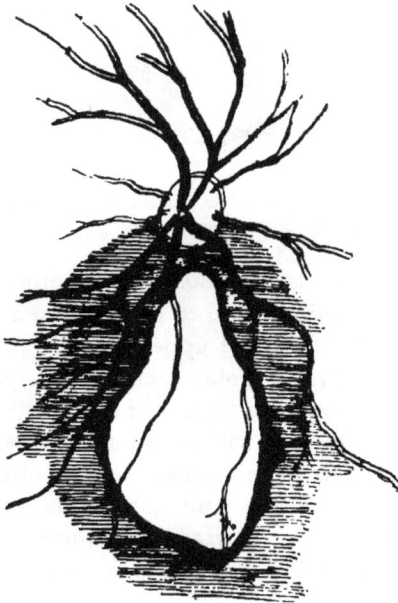

FIG. 121.—Congenital coloboma of the choroid.

crystalline lens. According to its size and the excursions it makes during the movements of the eyeball, it may interfere considerably with vision.

The *optic nerve* is sometimes but partially developed or it is atrophic. Very frequently we find that the change of *medullated nerve fibres* into non-medullated ones takes place in the retina near the optic papilla, instead of taking place during the passage of the optic nerve through the sclerotic.

In rare cases congenital atrophic spots are found in the choroid, or a coloboma of the choroid alone. (See Fig. 121).

The congenital · malformations of the *crystalline lens* have already been spoken of.

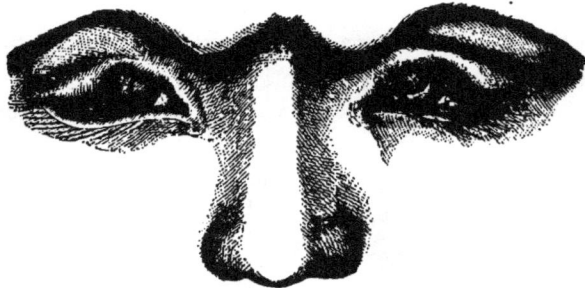

FIG. 122.—(After Manz). Congenital coloboma of the eyelids with adhesion between the upper eyelid and the cornea.

§156. *Coloboma* of the eyelids has sometimes been observed and has been usually found to be combined with *symblepharon*, a piece of skin from the upper lid being attached to the cornea. (See Fig. 122). Total lack of the eyelids, or a condition in which the skin passes uninterruptedly over the external orifice of the orbit and covers the usually deformed eyeball (*kryptophthalmus*), has been seen by several observers.

Congenital symblepharon and ankylopharon are rare. A more frequent affection is *ptosis* of the upper eyelid. In these cases the levator palpebræ superioris is either badly developed or totally wanting.

Epicanthus internus, a malformation which is often combined with ptosis of the upper lid and strabismus convergence (See Fig. 123), is usually found to affect both sides, and consists of a semilunar fold of skin, with its concavity towards the eyes, which connects the upper and lower lids at their nasal side.

The position of the lachrymal caruncle and lachrymal papillæ is not influenced by it. In rare cases epicanthus has been seen on the temporal side (*epicanthus externus*).

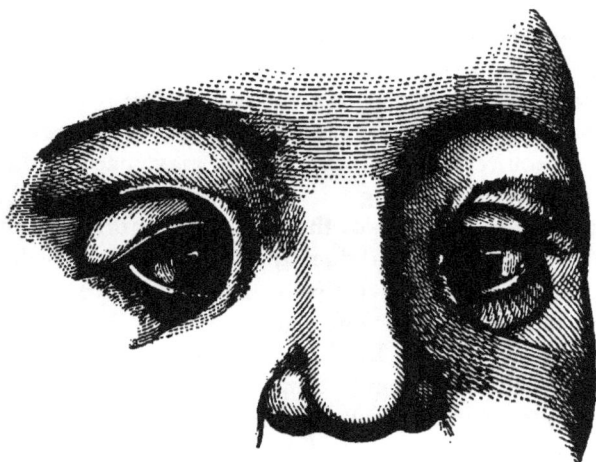

FIG. 123.—(After Von Ammon). Epicanthus internus with strabismus convergens.

The anomalies of the lachrymal apparatus are mostly absence of one or both lachrymal puncta or supernumerary puncta.

A number of *congenital tumors* are found in the adnexa of the eye. They are either *cysts* or *vascular tumors* of the orbital tissue and eyelids.

A congenital lack of pigment in the uveal tract, usually combined with lack of pigment of the hair, is known as *albinism*.

In rare cases the iris of one eye is differently colored from that of the fellow-eye, or we see in one and the same iris patches or sectors differing in color from the rest (*heterophthalmus, heterochromia*).

CHAPTER XIV.—EYE-AFFECTIONS DEPENDENT ON DISEASES OF OTHER ORGANS OR DISEASES OF THE GENERAL SYSTEM.

§157. *Respiratory apparatus.*—In affections of the respiratory apparatus, which cause a great deal of hard coughing, and especially in whooping cough, the rupture of a conjunctival bloodvessel is sometimes observed. The resultant ecchymosis may, of course, vary greatly in size. No special treatment is required. The same thing may happen during an attack of forcible sneezing.

Catarrhal inflammation of the mucous membrane of the nose may extend into the nasal duct and there cause a swelling and obstruction, and thus give rise to stillicidium (tear-dropping). If the case is an acute one, the symptoms in the tear-duct may pass away with it. In chronic catarrh of the nasal mucous membrane, the nasal duct becomes frequently permanently obstructed by scar-tissue, or by disease of the bones, due to the diathesis upon which the chronic nasal catarrh is based (syphilis and scrophulosis). In the treatment of these cases a careful examination of the nose must not be forgotten, and often the lachrymal trouble cannot be cured until the nasal mucous membrane is brought back to the normal condition.

Hypertrophic rhinitis, which is known to cause a number of reflex symptoms, may also produce reflex symptoms in the eyes, particularly a form of asthenopia which might be termed *nasal asthenopia.* It is, however, more frequently the cause of lachrymal conjunctivitis.

Ozæna may give rise to infectious ulcers of the cornea. Phyctaenular keratitis, blepharitis and follicular conjunctivitis may be due to infection from the nasal secretions.

Tumors of the nose, polypi of a benign or malignant character, are apt to encroach upon the orbit, and thus can cause exophthalmus.

§158. *Circulatory apparatus.*—Diminished arterial pressure, especially when caused by the insufficiency of the aortic valves, may cause pulsation of the arteries of the retina; this pulsation is isochronous with the systolic contraction of the heart.

If the force of the heart's action is considerably reduced, this may cause ischæmia of the optic nerve and retina. The heart in such cases is no longer able to overcome the normal intraocular tension, but it can do so again when the tension is reduced by paracentesis of the anterior chamber.

Pernicious progressive anæmia causes anæmia of the optic papilla and retina, accompanied by numerous hæmorrhages into the retinal tissue.

Hypertrophy of the left ventricle of the heart is apt to produce retinal hæmorrhages, or hæmorrhages into the vitreous body, and this may happen, whether the venous stasis is due originally to an affection of the lungs or of the heart.

Fibrinous endocarditis may be the cause of an embolism of the central retinal artery or of one of its branches.

Pulsating exophthalmus is a symptom which in almost all cases is due to an affection of the bloodvessels, and but very rarely to a pulsating tumor of the orbit. Its most prominent symptom is the exophthalmus, which is often one-sided, and may be very considerable. The upper eyelid is swollen, and its veins are dilated, and the conjunctiva shows dilated bloodvessels and serous infiltration. The pupil it generally dilated. The eyeball can be pressed into the orbit without causing pain, but it will protrude again as soon as the pressure is released. When the fingers are placed on the eyeball, it is felt to pulsate. Sometimes this pulsation is even visible. By auscultation pulsatory sounds are heard on the eye and the surrounding regions. Compression of the common carotid artery reduces these sounds or stops them altogether.

This exophthalmus may be accompanied by optic neuritis or œdema of the optic papilla. The retinal veins are always considerably enlarged and pulsate. Sight may be very much impaired, and in some cases blindness has been observed. The patients complain chiefly of the continued noise, and sometimes of pain. Paresis of the external muscles of the eyeball may cause diplopia. These symptoms have but very seldom come on without any known cause; in most cases they have developed after an injury, and for the most part after a heavy fall. They may follow rapidly upon the injury, which is the rule, or they may develop more slowly. When the exophthalmus occurs in both eyes, one eyeball usually has protruded before the other.

In a few cases the patients have been known to recover spontaneously from this affection, but in most cases, when not interfered with, death has been the result. The cases of idiopathic pulsating exophthalmus seem to be more frequent among women, while the traumatic ones have nearly all occurred in men (*Sattler*).

The anatomical cause of pulsating exophthalmus is sometimes an aneurysm of the ophthalmic artery, but in a large majority of the cases a rupture of the internal carotid within the cavernous sinus, with consequent effusion of arterial blood into this sinus and increase of blood pressure, causing dilatation of the superior ophthalmic vein, and ultimately of all the venous vessels communicating with the superior ophthalmic vein, and also of the inferior ophthalmic vein. The central retinal vein, which empties the blood either into the superior ophthalmic vein or directly into the cavernous sinus, soon shows therefore the same symptoms of dilatation and pulsation.

The therapeutic measures must, of course, be directed to the primary affection. As this is usually a rupture of the internal carotid artery, continuous digital compression or ligation of the common carotid artery must be performed.

We may here refer also to amblyopia, or amaurosis dependant on the *loss* of large quantities of *blood*, to whatever cause it may be due, such as wounds, ulcers of the stomach, the cancerous erosion of a larger bloodvessel, intestinal hæmorrhages, uterine affections, etc. If the patient recovers from the loss of blood, his sight, as a rule, will also gradually be regained.

§159. *Organs of Digestion.*—The congestion to the head caused by *chronic constipation*, and by hyperæmia of the liver, is sometimes combined with eye-symptoms. These are, in the main, an easily fatigued accommodation, and the appearance

of dark or light spots dancing before the eyes.   These symptoms disappear with the removal of their primary cause.

Leukæmia causes a form of retinitis which is said to be characterized by a yellow tint of the whole retina.

§160. *Uro-poetic Apparatus.*—Diseases of the kidneys give
rise to various eye affections. The affection which we call
albuminuric neuro-retinitis, and which has been already described in Chapter XIII, is generally due to the *shrinking kidney,* that form of nephritis in which the specific gravity of the
urine is usually low, and in which the albumen is small in
quantity, or may at times be wanting altogether.   Albuminuric retinitis is also found in acute croupous nephritis, as during scarlet fever, and in cases of amyloid degeneration of the
kidneys.   The disease of the kidneys may lead to albuminuric retinitis by causing an abnormal condition of the blood,
which brings on pathological changes in the coats of the blood-
vessels.   Further symptoms may be due to increased blood-
pressure and to the retention of urea.   Although this form of
retinitis belongs, as a rule, to the latter stages of the kidney
disease, it is sometimes the first symptom noticed by the patient; and its characteristic features may thus sometimes lead
us to detect a kidney disease when no other symptom is as yet
so pronounced as to suggest the diagnosis.

Another eye affection due to kidney disease is the so-called
uræmic amaurosis.   It is seen in all forms of nephritis, but
chiefly in acute nephritis from scarlet fever or the nephritis of
pregnant women.   The blindness usually comes on rapidly
during an uræmic attack.   The pupils are sometimes dilated.
Whether the patient has already been suffering from albuminuric retinitis or not, makes, of course no difference as regards
the occurrence of such a uræmic amaurosis.   The amaurosis
may pass off again after a few hours.   Sometimes, however, it
lasts even several days.

Albuminuric neuro-retinitis occurs, also, as a symptom of
nephritis of pregnancy.   It is then a very grave symptom and
an indication for bringing about premature labor.

§161. *Genital organs.*—Amblyopia is sometimes seen after

excesses *in venere.* It never lasts long in these cases, and does not give rise to a serious affection. *Onanism,* especially when frequently indulged in, causes the same symptoms. It · has become to be quite an accepted fact that onanism, as such may, and is likely to lead to serious eye affections, and it frequently happens that a patient comes to the physician frightened out of his wits after having read about the serious consequences of this habit. An examination either reveals no eye affection at all, or one which has nothing to do with the self-indulgence, and which in all probability has existed previously.

We may in this place speak also of a disease in which the eye-symptoms are very prominent, and which by many is thought to be due to an affection of the genital organs. I refer to exophthalmic goitre, or *Basedow's (Graves') disease.*

The cardinal symptoms of Basedow's disease are an increased action of the heart, exophthalmus of one or, generally, of both eyes, and goitre, although one of these three symptoms may be wanting.

The first symptom is generally the increased heart's action. The pulse ranges from 100 to 200 in the minute, the shock at the apex of the heart is felt much stronger than in the normal condition and can even be seen, although the heart is usually not hypertrophic. The carotid arteries and the veins of the neck pulsate visibly. Any excitement or tiresome work aggravates these symptoms. Sooner or later the thyrioid gland begins to swell. The tumor is at first soft, but later on it becomes harder in conseqence of the newformation of connective tissue or of deposits of lime within it. The swelling is at first not very large, and it may come and go. With the hand a tremor may be felt over the thyrioid gland, and with the stethoscope circulatory murmurs can be heard in it. The exophthalmus appears usually as the last of the three cardinal symptoms. The protrusion of the eyeballs is mostly in a forward direction, but sometimes there is divergent strabismus. In most cases the protrusion occurs in both eyes.

Combined with the exophthalmus is a loss of co-ordination in the movements of the eyeball and eyelids, so that in downward movements of the eyes the upper lids lag behind and expose a strip of the sclerotic. In the lower eyelids the same symptom is present, but it is less noticeable. This symptom (*Graefe's*) is considered as almost pathognomonic for this form of exophthalmus, and it is sometimes observed even before the exophthalmus has become very conspicuous. The palpebral fissure is generally very wide, and involuntary nictitation is wanting.

The pupils are often normal, in other cases dilated, the accommodation remains, however, undisturbed. The secretion of tears, at first increased, becomes later on diminished, and a conjunctival catarrh is seldom entirely wanting. The cornea becomes dry, since the eyelids do not protect it properly, and ulcerations, even the total destruction of the cornea, may be the result. The retinal arteries pulsate.

These cardinal symptoms are attended by a number of varying ones, as chlorosis or anæmia, uterine affections, higher temperature, etc. The palpitations of the heart cause dispnœa, orthopnœa, and even angina pectoris. The digestion is disturbed. Headache and insomnia are almost constant symptoms, and help to lower the vital power.

Basedow's disease rarely appears as an acute affection, and it usually takes several years for all its symptoms to become fully developed. Intermissions are the rule, and the characteristic symptoms of the disease may even exist for years, and then disappear. In about 12 per cent. of the cases (*von Dusch*) death results from exhaustion, from ascites, hæmorrhages into the brain, lungs or intestinal tract.

The diagnosis is easy when the cardinal symptoms are all well marked. In the beginning, however, the disease might be confounded with paralysis of the sympathetic nerve.

According to *Mooren, Basedow's* disease attacks twelve women to one man. It develops often after other severe, weakening diseases, great losses of blood, undue bodily exertion, etc. Sometimes it seems to be dependent on an hereditary tendency, and may show itself in several members of the same family.

Anatomical examinations have shown that the heart in *Basedow's* disease may be actually dilated and hypertrophied. The goitre is most frequently a glandular hypertrophy, but in some cases it has been found to be simply due to a dilatation of the bloodvessels of the thyrioid gland. The orbital fat is usually hypertrophic and œdematous, and its bloodvessels are dilated. The external muscles of the eye have been found in a state of fatty degeneration, but in a single case which I had occasion to examine, they were normal. Change of various kinds in the cervical part of the sympathetic nerve and its ganglia have been reported in some cases; in other cases no such changes could be detected. Pathological anatomy has, therefore, thus far revealed no common cause for all the symptoms observed in *Basedow's* disease, and it is only upon the clinical symptoms that any rational explanation can be founded.

The palpitations of the heart may be due to irritation of the excito-motory nerves of the heart, which arise from the medulla oblongata, enter the sympathetic nerve and leave it again with branches forming the cardiac plexus, or they may be due to paralysis of the inhibitory nerve-fibres coming from the vagus nerve.

Paralysis of the cervical part of the sympathetic nerve will account for the vascular symptoms in *Basedow's* disease, the goitre and even the exophthalmus; but it produces contraction of the pupil and of the palpebral fissure, while by irritation of the oculo-pupillary fibres of the cervical part of the sympathetic nerve, which spring from the anterior root of the second dorsal nerve, we can produce widening of the palpebral fissure, dilatation of the pupil and exophthalmus. We should, therefore, have to accept two different conditions in the tract of the same nerve in order to explain the more prominent spmptoms of *Basedow's* disease on the hypothesis of an origin in the sympathetic nerve.

It would be going too far to give here all theories which have been advanced in order to explain the symptoms of this disease. Suffice it to say, that paralysis of the center for the vagus nerve explains best the majority of the symptoms when combined with paralysis of a special (assumed) center for the bloodvessels of the orbit and thyrioid gland, and of the reflex-center for the movements of the eyelids. From this fact it appears that the symptoms of *Basedow's* disease are due to a brain lesion and, in fact, *Filehne*, by cutting into the corpora restiformia of rabbits without injuring the fourth ventricle, after having first severed the sympathetic nerve, has produced the cardinal symptoms of this disease.

Tonic treatment of all sorts has been recommended in this affection. The most successful treatment seems to be galvanization, combined with strophantus, or arsenic and iron internally.

If the cornea is endangered by ulceration, the palpebral fissure may be shortened by tarsorraphy.

§162. *Affections of the skin.*—*Erysipelas* of the face has in several cases led to thrombosis of the central retinal vein and atrophy of the optic nerve through compression, consequent on the extension of the inflammation to the orbital tissues.

Herpes zoster is sometimes found together with herpes of the cornea, and the latter is then considered to be due to a disease of the *ganglion Gasseri*.

§163. *Infectious Diseases.*—*Measles* in the eruptive stage give rise to conjunctivitis. Chronic conjunctivitis conjoined with blepharitis ciliaris, and phlyctænulæ, and even parenchymatous keratitis, are often seen after measles, and their occurrence may be due to the lowering of the whole system by this disease, or to direct infection. Optic neuritis and amaurosis have also been observed after measles, but they are extremely rare.

Scarlet fever, by causing nephritis, may bring about an albuminuric neuro-retinitis, or a uræmic amaurosis. *Diphtheria* of the conjunctiva may be seen in combination with scarlet fever.

Small-pox may give rise to a great variety of eye-affections, not counting, of course, the fact that pustules may be located on the skin of the eyelids.

Catarrhal, purulent, and diphtheritic conjunctivitis, keratitis, blennorrhœa of the lachrymal sack, iritis and choroiditis are often observed during and after this disease. The most frequent affections, however, are those of the conjunctiva and cornea. The corneal troubles have also been observed to develop some weeks after recovery from the small-pox. They are

usually ulcerations and parenchymatous inflammations, which
lead to the formation of scars, or even to total destruction of
the eyeball.   Many an eye which has been lost in this manner
might have been saved, had the eye-affection been treated in
its earlier stages, and as the treatment is in most cases simply
local, and in no way interfering with the treatment of the
small-pox, there is no good excuse for neglecting it, even during the active period of the general disease.

Typhus abdominalis may give rise to corneal abscesses or
ulcerations, paresis of the external muscles of the eyeball or
of the accommodative apparatus, or to orbital cellulitis.

Amblyopia and amaurosis are sometimes seen after typhus,
as they are after other prostrating diseases, and are due to an
anæmic condition of the optic nerves.   In nearly all of these
cases sight returns with the improvement of the condition of
the general system.   Sometimes atrophy of the optic nerve is
observed.

Febris recurrens is said to cause by preference affections of
the uveal tract.

Diphtheritis of the throat appears very seldom to cause *diphtheritic conjunctivitis,* but in some rare cases the disease reaches
the conjunctiva through the lachrymal passages. A much more
frequent affection following diphtheritis of the throat is paresis of the accommodation, already referred to in Chapter XX.
The physician should at once suspect it, when some weeks
after recovery from the diphtheritic attack in the throat, vision
for near objects becomes weakened or imperfect.

Malarial fever, as has been stated, is thought by many phy-
sicians in the Mississippi Valley to cause all sorts of eye-affections.   There is no question that chronic conjunctivitis and
trachoma, are very frequent in the fever districts, yet, I have
not seen that quinine has had any beneficial influence in pro-
moting their cure. Malarial keratitis has been described (*Kipp*)
as a special form of keratitis.   I have sometimes seen small,
point-like infiltrations of the cornea, which have appeared in
connection with malarial fever. Malarial optic neuritis and
œdema of the optic papilla have been described by *Macnamara.* Paralysis of one or more of the external muscles of the
eyeball may also be due to malarial poisoning.

Cerebro-spinal meningitis may cause partial or total atrophy of one or both optic nerves by constriction of the nerves at the base of the brain. In some cases it produces purulent panophthalmitis.

The recent epidemic of *influenza* (*grippe*) has shown this disease capable of bringing about numerous forms of eye diseases. Abscesses of the lids, very severe acute conjunctivitis, phlegmome of the orbit, paralysis of ocular muscles, optic neuritis, atrophy of the optic nerve, iritis and severe cases of purulent panophthalmitis, have all come under my own observation as sequelæ of influenza.

Tuberculosis may affect the conjunctiva, and is found in the uveal tract. It may be the primary form of infection, or the ocular tuberculosis is due to the dissemination from some other center of infection.

Pyæmia and septicæmia cause eye-symptoms, which may be due to the thrombosis of the cavernous sinus, or of one of the ophthalmic veins. In other cases, especially in puerperal septicæmia, purulent choroiditis (*choroiditis metastatica*), due to microbic embolism, has been observed. I have seen it also follow a purulent arthritis. A septic retinitis has also been observed and described.

Acquired syphilis shows itself in the eyes in a great many ways, and in all periods of the disease. Primary chancres have been met with on the skin of the lids and on the conjunctiva. The part of the eye most frequently attacked by syphilis is the uveal tract. The commonest form of syphilitic eye-disease is iritis. It is usually a simple plastic iritis, which appears at the same time with the skin symptoms, or at a later period when other syphilitic symptoms are no longer recognizable. In some cases the iritis is of a recurring type. Such a plastic syphilitic iritis may begin rather quietly and with but little pain, and may thus differ somewhat from certain other forms of iritis; yet, as a rule, there is no symptom which absolutely proves an iritis to be of syphilitic origin, unless it be the formation of a gumma. Gumma of the iris is easily recognized, as has been described in Chapter X. The gumma may remain small, or it may gradually increase in size so as to nearly fill the anterior chamber.

Syphilitic choroiditis may appear as disseminate choroiditis or as a central chorio-retinitis, or it may be an exudative choroiditis. Gummata have also been observed in the choroid. Syphilitic choroiditis is found in patients of a more advanced age, and occurs mostly at the same time with or soon after the so-called secondary symptoms, or at a very late period.

Syphilitic retinitis, usually conjoined with choroiditis, and syphilitic neuritis, are sometimes met with; also cyclitis, gummata of the ciliary body and of the sclerotic. Atrophy of the optic nerve and paralytic symptoms in the external muscles of the eyeball are often due to syphilis. Sometimes it causes diseases of the lachrymal apparatus and chronic hyperæmia of the conjunctiva bulbi.

Hereditary syphilis is often the cause of parenchymatous keratitis, and sometimes of iritis and choroiditis. Such a keratitis is also occasionally caused by acquired syphilis, and I have seen it follow slight injuries to the cornea in syphilitic subjects.

§164. *Intoxications.—Lead-poisoning* is apt to cause optic neuritis, transient amaurosis or even atrophy of the optic nerve. The eye-affection generally precedes the general symptoms.

Progressive atrophy and central scotoma due to *alcohol* and tobacco intoxication, have been detailed in Chapter XIV.

Toxic effects from eating foul sausage, meat-pastry or fish (pike), have in rare cases produced a paresis of the accommodation, exactly like that observed after diphtheritis. Sometimes it has been conjoined with amblyopia.

Intoxication with belladonna, hyoscyamus, datura, duboisia and gelsemium, causes dilatation of the pupil and paralysis of the accommodation.

Morphia and opium intoxication, in the acute forms, causes miosis of the pupil.

Quinine intoxication has especially of late been found to cause amblyopia and amaurosis. The latter may become permanent, or some small area of the field or even central vision may be re-established. The ophthalmoscope shows anæmia of the optic nerve and retina. The affection usually leaves an impairment of the color-sense and the light sense behind even if vision is regained.

§165. *Diabetes.*—*Diabetes mellitus* is sometimes the cause of the formation of cataract. While some operators are afraid to extract such cataracts, and prefer to use the suction method, others do not acknowledge any special danger from operations for diabetic cataract. I have never seen any disagreeable accident following extraction in such cases.

The optic nerve and retina are sometimes found to be inflamed in diabetes mellitus, and the ophthalmoscopic picture is similar to that of albuminuric neuro-retinitis. Furthermore, amblyopia and atrophy of the optic nerve and paralysis of the external muscles of the eyeball have been found, caused apparently by diabetes.

Rheumatism may give rise to iritis, and it is especially the chronic relapsing form of iritis which is usually ascribed to it. Some forms of paralysis of the external muscles of the eye and of episcleritis, for which no other cause can be found, are conventionally termed rheumatic.

Iritis is also often due to a *gouty diathesis*, and sometimes cyclitis is dependent on the same general disease.

Scrophulosis.—*Scrophulosis, strumous habit,* is the most frequent cause of phlyctænular affections of the eye and sometimes of parenchymatous keratitis, chronic catarrhal conjunctivitis, blepharitis and affections of the lachrymal apparatus.

CHAPTER XXV.—ON THE DETECTION OF ONE-SIDED SIMULATED BLINDNESS AND CONGENITAL COLOR-BLINDNESS.

§166. The oldest and simplest method for the detection of *simulated one-sided blindness* consists in placing a prism before the eye which is pronounced to be healthy, while the individual is looking at a distant object, and thus to evoke double vision. It is best to hold the prism before that eye with its base upwards or downwards. If the individual under examination acknowledges his diplopia, the eye pronounced blind must necessarily see, and his binocular vision is demonstrated.

In many cases, however, the malingerer is acquainted with this method, and it must then be modified. He is directed again to look at a distant object; then the examiner covers the so-called blind eye, and holds a prism before the so-called good eye in such a manner that the prism covers only about half the pupillary space, thus producing a monocular diplopia. If the malingerer does not acknowledge this diplopia, he is to be suspected. If he acknowledges it, we proceed to uncover the so-called blind eye, and shift the prism so that it covers the whole pupil, thus changing the monocular into a binocular diplopia. He must, of course, not be allowed to suspect the trick, and if he continues to 'see double, the so-called blind eye must see. If the prism be used in the same manner, and no eye is covered, three images must appear, and this method may be used in certain cases.

Another method is to let the malingerer read, and to exclude his so-called good eye from sight by some means while he is reading. This may best be done by holding a very strong convex glass before it, as but few malingerers will be stupid enough to go on reading, when a dark or ground glass is held before their so-called good eye.

If, further, the malingerer goes on reading undisturbed and

without shifting the book or his head, when a pencil or some such object is held between his eyes and the book, he must see with both eyes.

Another excellent test (*Snellen*) makes use of red and green letters which the patient is made to read through red and green glasses. If the green glass is held before the good eye he can only see the green letters, or if the red glass is held before it he can only see the red letters. Of course if, under the circumstances, he reads all the letters, he must see with both eyes.

The so-called good eye may also be excluded from vision by the instillation of a mydriatic. However, but few malingerers will allow anything to be put into their so-called good eye.

§167. *Congenital color-blindness.*—Color-blindness has of late become an important subject in certain branches of modern civilization. The fact that a man is color-blind, evidently unfits him for any service in which the prompt fulfillment of important duties depends on his recognizing colored signals.

It has, therefore, become a law in most civilized countries that men applying for positions in the railroad service, or in the marine, must first undergo an examination with regard to their color-perception.

Color-blindness may be total or only partial. In total color-blindness the patient perceives only black and white, and all other colors are to him either white or black or some intermediate shade of gray. In partial color-blindness the patients generally see two complementary colors besides white and black. In the so-called red-green blindness yellow and blue are perceived; in the so-called blue-yellow blindness red and green are recognized. These two forms are the typical ones of partial color-blindness, yet slight variations are often observed. The visual acuteness of eyes which are color-blind is generally perfectly normal.

Various methods have been devised in order to detect partial congenital color-blindness. The simplest in common use is that of *Holmgren.* The patient is given a skein of colored worsted, and directed to select from a bundle containing all sorts of colored worsted those skeins which appear to him of the same color as the given one (usually at first a pale green or a pale pink). If there is any hesitation in matching the color, or if he selects different colors to match the given one, his color-perception cannot be normal.

If, for instance, he matches a light green skein with red, brown or gray, he is surely color-blind. If he is red-green blind, he will mix the colors up on the principle that to him blue and yellow only are distinct colors, and all other colors appear to him as shades of yellow (*Mauthner*). If he is, however, blue-yellow blind, he will see only red and green as distinct colors, and every other color will appear to him as a shade of red.

Red-green blindness is by far the commonest form of partial color-blindness.

The affection is but very rarely monocular, and nothing definite is known with regard to its etiology, although it is most likely due to a lack of development of the center for color-perception.

Acquired color-blindness, associated with progressive atrophy of the optic nerve, has been detailed in Chapter XIV.

CHAPTER XXVI.—ASEPSIS AND ANTISEPSIS IN OPHTHALMIC SURGERY. DESCRIPTION OF THE MOST IMPORTANT OPERATIONS ON THE EYE AND THE EYELIDS.

§168. Whether a surgeon be a believer in the theory of the microbian origin of disease or not he cannot, at this day, afford to disregard the methods of *asepsis and antisepsis,* by which surgery has made such undoubtedly great strides towards perfection. In ophthalmic surgery in this regard the same rules hold good as do elsewhere. We must strive to be aseptic, to operate with aseptic instruments, and to render as far as possible aseptic, and keep it so, the field of our operative interference. Where asepsis cannot be produced, antisepsis must take its place. How the surgeon is to render himself and the surroundings of the patient sufficiently aseptic (the ideal can never be reached) we can not detail here. The instruments are best made aseptic by being placed in boiling water or by sterilizing them by means of steam. When this cannot be done they should be placed into a two per cent. solution of carbolic acid or creoline. As both these solutions are apt to affect the edge of cutting instruments disagreeably, the latter are best disinfected by means of absolute alcohol. The neighborhood of the eye, particularly the eyebrows and eyelashes and their roots are best disinfected by first scrubbing and washing them with soap and then with a solution of bichloride of mercury of one part in four thousand. It is well in order to allow the bichloride of mercury solution to remain in better contact with the parts to be disinfected, to cover them for some time just before the operation with a layer of cotton saturated with it. The conjunctival sack must be repeatedly flushed with the same solution just before operating. Other solutions or powders are used by others to reach the same end (boracic acid, carboli acid, iodoform, aristol, etc.). All drugs in solution dropped into the eye in connection with an operation, as cocaine, eserine and atropine, must be sterilized if possible, by heating. Cocaine and atropine should be dissolved in a four per cent. solution of boracic acid, instead of distilled water alone. Scrupulous asepsis and antisepsis are to be carried out by these means or others. When in spite of all precautions an infection of the operative wound takes place, the last resort is cauterization by means of the actual or galvano-cautery. If there is any stoppage in the lachrymal drainage apparatus, or any infectious disease of the conjunctiva present, it must be rendered innocuous before an operation on the eyeball is performed. The method of bandaging, which I consider the ideal one, has been detailed in Chapter VI.

§169. General anæsthesia is but seldom required in operations on the eyeball, except in children, since *Koller's* discovery has given us the local anæsthetic, cocaine. It is best to use this in a two to four per cent. solution and to instill it into the conjunctival sack several times at short intervals (from three to five minutes) before an operation on the eyeball is performed. It may also with advantage be injected under the skin of the lids and into the deeper tissues of the orbit in certain operations, but the pain it causes and the possibility of poisonous effects are serious objections against this manner of causing local anæsthesia. Tropa-cocaine is now recommended as non-poisonous and less irritating, in the place of cocaine, particularly for injections.

The ophthalmic surgeon should be ambidexter, that is, he should have trained his left hand to be as useful and free in action, or nearly so, as his right hand.

§170. *Tenotomy.*—The tenotomy of one or more of the external muscles of the eyeball is performed for the correction of strabismus or insufficiency. The operation is most frequently done on the internal rectus, more rarely on the external rectus and others.

When possible, it is best to perform this operation without

putting the patient under the influence of a general anæsthetic because the effect of the tenotomy can then be promptly estimated, and, if necessary, be improved upon. By the help of cocaine this can be easily done, even in comparatively small children.

After the eyelids have been separated and are held apart by means of a wire-speculum, or by the fingers of an assistant, the patient is ordered to roll the eyeball in the direction opposite to the muscle to be cut. The conjunctiva and episcleral tissues are then firmly grasped with strong toothed forceps below the insertion of the muscle and somewhat nearer the corneo-scleral margin. The fold of tissue thus grasped is then cut by means of strabismus scissors. If the first clip has not severed *Tenon's* capsule, this must be done by a second cut. When this is accomplished, but not before, the strabismus hook is slipped under the tendon of the muscle through this external incision. The handle of the hook is then raised so as to put the muscle upon the stretch, the strabismus scissors are introduced and the tendon is cut close to the sclerotic. By then bringing a second strabismus hook behind the first one, entering it with its point downwards and sweeping the sclerotic with it while turning the point upwards, or *vice versa*, any stray tendinous fibres are detected, and are to be severed. Only when the strabismus hook can be moved under the conjunctiva close up to the corneo-scleral margin, without encountering any obstacle, we may be sure that the muscle is perfectly divided.

When this is done, the effect of the operation should be tested. If the motility of the eyeball in the direction of the muscle operated upon is greatly reduced, the desired effect is probably attained. If the internal rectus has been cut, the patient ought to be able to move the inner margin of the cornea as far inward as the lachrymal caruncle, and if he can not do this, the effect of the tenotomy may be reduced by drawing the eyeball by a suture to the inner angle of the palpebral fissure (*Knapp*). If the patient can move his eye farther toward the nose, the tendon is not perfectly divided, and the hook and scissors must be re-introduced.

If tenotomy has been performed on the external rectus, the

patient should still be able to move the outer margin of the
cornea as far as the outer angle of the palpebral fissure; and
if he cannot do so, the effect of the operation may be reduced
by drawing the eyeball to the outer angle of the palpebral
fissure by means of a suture.   To increase the effect of a tenotomy a suture must be placed in the opposite angle of the
palpebral fissure.

Tenotomy has generally to be performed on both eyes in
order to gain a perfect result.   This must, however, never be
done at one sitting, when the tenotomy is made to cure convergent strabismus, since an over-correction may result in a divergent strabismus in place of the convergent one, for which
the operation has been performed.

Partial tenotomies which are said to enable the surgeon better to grade the effect of the operation, have been introduced
as substitutes for total tenotomy of the antagonist of an insufficient muscle (*Stevens*).   Instead of severing the tendon
of such a muscle a piece is cut out of it by specially devised small instruments in the hope of weakening the antagonist just to the degree required to establish the equilibrium
between the two muscles.   The results, from what I have seen,
seem very doubtful.

§171.   *Advancement* of an ocular muscle may be performed
alone or combined with tenotomy of the antagonist.   While
tenotomy attempts to render a too strong muscle weaker by
causing its insertion to be moved backwards on the eyeball,
advancement renders the weak muscle more effective in its action by shortening it, or by bringing about an insertion closer
to the cornea, or by both measures together.

By some operators this latter method of correcting strabismus is always considered to be the preferable one.   Its application is commonest, however, in divergent strabismus, as a
simple tenotomy of the external rectus or recti in this affection
is but seldom successful.   It is, therefore, necessary in these
cases to advance the insertion of the internal rectus to a position nearer the corneo-scleral margin, and thus in effect to
shorten the muscle.   The operation for the advancement of
the internal rectus must nearly always be combined with the

division of the external recti, and in some cases it may be necessary to perform the operation for advancement on both internal recti.

The conjunctiva is first incised over the insertion of the internal rectus in a vertical direction, and the muscle is grasped with the forceps, or a thread is drawn through it, so as to prevent it from slipping backwards and out of reach when the tendon is cut. This cutting of the tendon is done close to the sclerotic, and the muscle may, if necessary, be shortened to the required degree by cutting off a piece. When this is done, the portion of the conjunctiva which lies between the incision and the corneal margin is indermined with fine scissors, and the internal rectus is drawn under it, and fastened in this position by means of two or three sutures. (See Fig. 124).

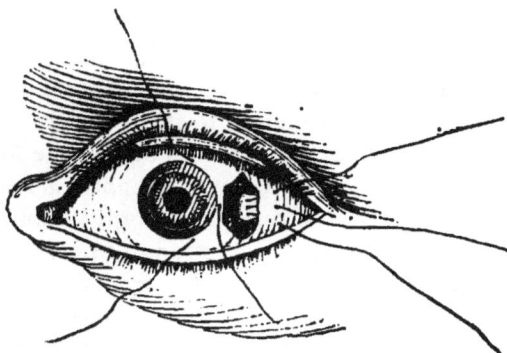

FIG. 124.—Advancement of the rectus externus.

The required effect is only reached when immediately after the operation, the eyes show a slight degree of convergence. This apparent over-correction disappears during the process of healing.

A more modern procedure introduced by *De Wecker* is the advancement of *Tenon's* capsule by means of a suture, which causes a fold in the muscle. When this fold has become adherent to the underlying part of the muscle, the muscle is practically shortened and accordingly stronger.

§172. *Enucleation of the eyeball.*—The operation of remov-

ing an eyeball by enucleation is in most cases best performed
while the patient is under the influence of a general anæsthetic.
When necessary it can, however, be done under local cocaine
anæsthesia.

The eyelids are held apart by a wire speculum or by the fingers of an assistant, and the conjunctiva at the periphery of the
cornea and near the insertion of one of the recti muscles, preferably the inferior one, is grasped with the toothed forceps
and freely incised. A strabismus hook is then inserted under
the muscle, and all the tissue which can be lifted up by the hook
is divided with the scissors close to the eyeball, always cutting the conjunctiva first at the corneo-scleral margin. When
all the tissue which has been lifted by the first hook is severed,
a second hook is inserted behind the first one, and the tissues
lifted by this one are cut in turn, and so on around the periphery of the cornea. When the tendons of the inferior, external and superior rectus have thus been detached, it is best to
divide the internal rectus somewhat further back from its insertion so that the stump adhering to the sclerotic can be used
in order to rotate the eyeball outwards during the final step of
the operation, which is the cutting of the optic nerve. To accomplish this the eyeball is turned strongly outwards (some
operators prefer to turn it inwards, strange to say), and a large
strongly curved pair of scissors is introduced, with blades
closed, at the nasal side between the loosened orbital tissue
and the eyeball, and is pushed backwards until it has reached
the posterior surface of the eyeball. By moving the point of
the scissors up and down, the resistence felt will now enable
the operator to make sure of the position of the optic nerve.
When this is ascertained, the scissors are slightly withdrawn,
opened and advanced again, so as to catch the optic nerve between the blades. The optic nerve and the ciliary nerves
around it are then divided by one clip of the scissors, and the
eyeball is lifted out of the orbit with the scissors. The oblique
muscles and any further adhesions are now cut as quickly as
possible, since a comparatively profuse hæmorrhage takes
place as soon as the optic nerve is severed.

§173. *Paracentesis of the cornea.*—Paracentesis (puncture) of

the cornea and emptying of the anterior chamber is performed
for the relief of an increased intraocular tension, or for the
removal of pus or other pathological contents from the anterior chamber.

This little operation is usually performed by means of a
needle, a stop-needle, or a small lance-shaped knife. In puncturing the cornea, great care must be taken not to wound the
iris or the crystalline lens, as the former may lead to hæmorrhage or iritis and the latter will cause a cataract to develop.

§174. *Abscision of a total corneal staphyloma.*—Corneal
staphylomata are, as a rule, cut off by detaching the lower or
upper half of the protrusion with a *Beer's* knife, and the remaining half with scissors.

If the crystalline lens, which is usually cataractous, is still in
in situ, it should be removed after opening the anterior lens
capsule.

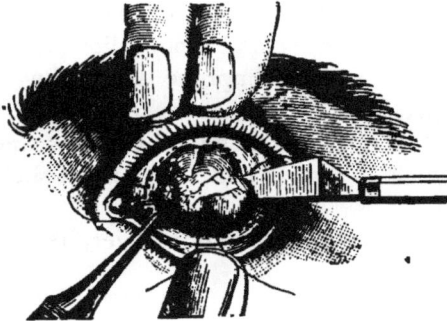

FIG. 125.—Removal of a total staphyloma of the cornea. The sutures are *in situ*
according to the method of Knapp.

The wound resulting from this operation may be left to heal
without further interference, or it may be closed by sutures.
The sutures must be inserted, ready to be tied, before the
staphyloma is removed. This is generally best done by *Knapp's*
method. A needle armed with a long thread, is entered under
the conjunctiva and episclera above the upper corneo-scleral
margin, and a little to one side of the vertical meridian, and
is brought out somewhat above the horizontal meridian on the

same side; it is then entered again a little below this meridian and brought out at a point below the lower corneo-scleral margin corresponding to the first point of entrance. (See Fig. 125). The same procedure is then repeated on the other side of the vertical meridian. After the removal of the staphyloma the threads are tied and the wound is closed. By this means an excellent stump for the wearing of an artificial eye may be obtained.

§175. *Evisceration* has been of late introduced as a substitute for enucleation for various reasons (*A.Graefe*), chief among which is the erroneous idea that removing a panophthalmitic eye may cause meningitis. The sclerotic being incised concentrically with the periphery of the cornea and near it with a knife, the circular section is completed by means of scissors. The cornea and a ring of sclerotic tissue attached to it are thus cut off. The contents of the eyeball are then removed and the inner surface of the scleral shell is scraped clean with a sharp spoon. The conjunctival and scleral wound lips may now be closed by sutures. In order to improve upon the stump resulting from this operation, *Mules* has devised an artificial vitreous body, a ball of glass or some other light and uncorrodable material, which is inserted into the scleral shell before closing the wound.

The healing after this operation is very painful and protracted. Its value is doubtful.

FIG. 126.—Scleral incision leaving a bridge upwards. Von Wecker's sclerotomy.

§176. *Sclerotomy.*—The operation of sclerotomy is performed to relieve an increased intraocular tension, and has of late been especially recommended as a means of curing certain forms of glaucoma. The operation is generally performed with a *Graefe's* cataract knife. This knife is entered in the corneo-scleral margin just in front of the insertion of the iris, and is brought out on the opposite side of the cornea, in the corneo-scleral tissue. The section may now be finished, and thus a corneo-scleral flap be formed (See Fig. 126), or, what is better, because it is less likely to be followed by prolapse and subsequent incarceration of the iris, a narrow bridge of corneo-scleral tissue is left uncut (*De Wecker*). The knife is then slowly withdrawn and eserine instilled.

§177. *Iridectomy.*—The operation of iridectomy, which consists in the removal of a sector of iris-tissue, is one of the operations most frequently performed upon the eyeball. It is the ordinary operation for an artificial pupil to restore sight to an otherwise useless eye, and for the relief of increased intra-ocular tension in cases of primary or secondary glaucoma; it may also make a part of an operation for the extraction of cataract, or the removal of a foreign body lodged in the eye.

The cornea is incised by means of a lance-shape knife or a *Graefe's* cataract knife in the corneo-scleral tissue, or nearer its center when a very small pupil is desired. When the knife is withdrawn, the iris may at once prolapse, in which case it is easily grasped by a pair of iris forceps, is gently pulled out and cut off as close to its ciliary insertion as possible. If the iris does not prolapse, the forceps must be introduced through the corneal incision, in order to grasp and draw it out. (See Fig. 127).

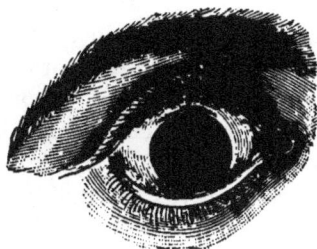

Fig. 127.—Scleral incision for iridectomy and appearance of the pupil after completion of the iridectomy.

Great care must be taken during this operation not to wound the crystalline lens, as this would subsequently cause the formation of cataract; also that no iris-tissue is allowed to remain lying between the lips of the corneal incision, since this might give rise to the formation of an ectatic scar and later infection through it.

In eyes in which vision has become abolished in consequence of a plastic iritis with occlusion of the pupil, it is sometimes very desirable but not at all easy to re-establish some kind of a pupillary opening. It is generally necessary to use iris forceps with a different arrangement of the teeth or a blunt hook (*Tyrell's*) in order to get hold of the iris. The variations in this little operation are almost as numerous as the cases on which it has to be performed.

§178. *Iridotomy.*—The operation called iridotomy or iritomy (*De Wecker*) is now often executed with the iridotomy scissors introduced by *De Wecker*. Its usefulness is most marked in those cases in which after a cataract extraction a secondary membranous cataract has been formed, or where in consequence of irido-cyclitis (especially after an injury with loss of the crystalline lens) the iris and cyclitic membrane together have formed a diaphragm closing the pupil. The presence of this diaphragm obliterates vision, although light-perception and projection, perhaps, may be very good. Such a diaphragm presents, moreover, a serious obstacle to the current of the intraocular fluids.

After the cornea has been incised by means of a lance-knife or *Graefe* knife to such an extent that the iridotomy-scissors can be easily introduced into the anterior chamber, the scissors are entered through the corneal incision and their sharp and pointed blade is thrust through the diaphragm. Then by closing the scissors, a cut is made without appreciable dragging on the ciliary body.

If successfully accomplished, the divided tissues retract, open out the slit, and thus a pupil is established. Sometimes it is necessary to make a second cut at an acute angle with the first and thus to isolate a triangular piece of tissue, the apex of which will generally either curl up or become retracted, and so give a permanent opening, or it may be pulled out of the eye by means of iris forceps and be cut off.

Iridotomy may also be performed in various other ways, as for instance, with a sharp narrow knife, like Graefe's cataract knife, *Culbertson's* iritome, or better with a knife-needle. When using a knife-needle it is also possible to avoid dragging upon the ciliary body, and we dispense with bringing so large an instrument as the iridotomy-scissors, or the iritome into the eyeball.

§178. *Extraction of cataract.* What is usually described as simple linear extraction is applicable particularly to soft and traumatic cataracts. (See Fig. 128). In such cases a small incision is made into the cornea near the point corresponding to the edge of a dilated pupil. If the lens capsule has previously not, or insufficiently, been opened, this is now done by means of a cystotome. The soft lens substance is then easily squeezed out through the corneal wound. In a similar manner shrunken and membranous cataracts may sometimes be drawn out of the eye with a small hook.

FIG. 128.—Corneal incision in simple linear extraction of cataract.

For the extraction of senile cataracts a number of procedures are in use. They differ from each other particularly in the position of the corneal incision and may be combined with an iridectomy or be done without it, or when an iridectomy is made, this may differ in size.

Daviel's method, improved by *Beer,* was to make a large corneal flap, the incision lying in the lower periphery and comprising about one-half of the circumference of the cornea. Then the anterior lens capsule was opened by a small hook or needle and the lens squeezed out.

This operation left the iris untouched, and when the wound healed without an accident, a round moveable pupil was left behind. This method was, for a long time, the generally adopted one. As asepsis and antisepsis were unknown, and local anæsthesia not in existence, a comparatively large per-centage of eyes so operated upon were lost by suppuration and by accidents during the operation.

Von Graefe, convinced that wounds in the sclerotic tissue would heal easier and when made linear (in the line of a great-est circle) would not gape, but would close readily, introduced the section which lay totally in the tissue of the corneo-scle-ral margin. (See Fig. 129). To be able to make a section large enough to allow of the passage of a very large lens he replaced the knives hitherto in use (*Beer's* triangular knife and the lance-knife) by the narrow bladed straight knife which bears his name. He, furthermore, added an iridectomy, which not only rendered the exit of the lens easier, but was also thought to prevent prolapse of the iris with its disagreeable consequences (*modified von Graefe extraction*). In finishing the scleral section a conjunctival flap could be made. Then followed opening of the lens capsule and squeezing out of the cataract.

FIG. 129.—The ideal corneo-scleral incision in Von Graefe's method of extracting cataract.

The healing after this method, which in turn was slightly modified by most operators so as to bring the height of the section into the corneal tissue, was considerably better than after the old flap extraction. Suppuration of the wound occurred in comparison but rarely. It was, therefore, until recently, and with some still is, the only method of extracting a cat-aract.

Aseptic and antiseptic methods and the local anæsthesia produced by cocaine, made the operation more free from acci-dents during its performance and afterwards. The aseptic and antiseptic measures were, further, supplemented by the washing out of the lens capsule and anterior chamber after the extraction of the lens (*McKeown*), and some surgeons have adopted this maneouvre as a routine performance.

The chief blemish of *von Graefe's* modified extraction is the iridectomy, which mames the eye, and by allowing an undue amount of diffuse light to enter an almost immoveable pupil often reduces the visual acuity.

FIG. 130.—Corneal incision (corneal flap) in simple extraction of cataract.

Most ophthalmic surgeons, therefore (See Fig. 130), have now returned to an operation without iridectomy and make, in consequence, a corneal flap, comprising about one-third of the circumference of the cornea. (See Fig. 131). After this

FIG. 131.—Modern simple extraction of cataract (corneal flap, no iridectomy).

follows the opening of the lens capsule and squeezing out of the lens substance. This in turn is followed by replacing the iris and the instillation of eserine. This method is now called *simple extraction.* The advantages it offers and the ease with which healing takes place in these days of asepsis and antisepsis, 'make it the method of cataract extraction of the future.

§180. *Discission of the anterior lens-capsule.*—In order to bring the lens-substance in contact with the aqueous humor and to thus cause its gradual dissolution and absorption within the eyeball, we must divide the anterior lens-capsule and the lens-substance. This may be done in most cases of cataract in persons under the age of 30 years. The little operation is generally executed with a needle, and it must usually be repeated several times before a perfect success is reached. Care must be taken not to wound the iris, and to make the first division of the capsule very small, as the too rapid swelling of the lens-substance may give rise to glaucomatous symptoms.

A similar operation has to be performed after the extraction of a cataract when the lens-capsule remains clouded (secondary cataract), and so interferes with the perfect restoration of vision.

§181. *Pterygium operations.* To remove a small pterygium successfully it is usually sufficient to separate it carefully from the cornea and sclerotic, and to excise with it a little of the conjunctival tissue so as to leave a wound of an approximately rhomboidal shape. The conjunctiva can then be stitched together to cover the denuded sclerotic.

For large pterygia this simple little operation is also perfectly as successful as others, when the wound is cauterized (I use pure carbolic acid) after the dissection and the conjunctival sack is flushed subsequently for some weeks several times a day by a solution of bichloride of mercury. I have of late discarded the stitching of the conjunctival wound lips altogether, but usually incise the conjunctiva near the cornea adjoining the site of the pterygium in such a manner as to allow it to withdraw from the cornea. By this means the corneal wound is healed over before the conjunctival loss of substance is replaced, and the result is better.

Prince recommends to tear the pterygium from its corneal attachment, and claims that the cornea is finally clearer than when abscision has been made.

For large pterygia several other methods are in use. The whole pterygium, after having been cut off the cornea and sclerotic, can be transplanted into the fornix of the conjunctiva where an incision has been made to receive it (*Desmarres*). This has been improved upon by *Knapp* who, after having severed the pterygium from the cornea and sclerotic down to its base, cuts it into two halves from point to base. The incisions into the conjunctiva are then carried further upwards and downwards into the cul-de-sac, and when this is done, each half of the pterygium is stitched into the corresponding gap resulting from these incisions. The operation yields lasting results.

A simple method which has been in use with some operators for many years (*Galezowsky*), and has found its warm advocates, has given me also perfect satisfaction. It is done in the following way: After the pterygium has been dissected from the cornea and sclerotic, down to its base, the conjunctiva at its base is undermined, so that the whole pterygium can be doubled upon itself and folded under it. It is then sewed to the conjunctiva in this position, and finally atrophies.

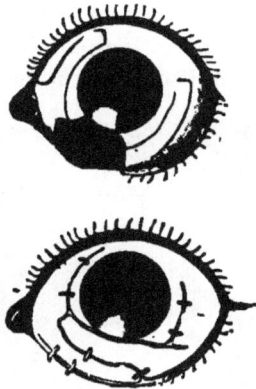

FIG. 132.—Teale's method of making conjunctival flaps for the covering of the ocular wound-surface resulting from the severing of a symblepharon. In the lower picture the flaps are in position, as at the completion of the operation.

§182. *Operations for the cure of symblepharon.* A small bridge-like symblepharon can be simply divided with the scissors, but if the union between the eyelid and the eyeball reaches far down into the fornix, usually more must be done to prevent the large wound-surfaces from growing together again. This can be accomplished by transplanting flaps of the conjunctival tissue of the same eye upon the ocular wound surface (See Fig. 132), or by transplanting a conjunctival flap without pedicle from another human or animal's eye upon one or both of the two wound-surfaces. Cutaneous flaps have also been successfully made use of. They are taken from a distant part (arm, *Kuhnt*) and without a pedicle. They may also be taken from the eyelid, in which case the skin is drawn through a button-hole-like aperture. made in the eyelid, and stitched to its its inner surfuce.

§183. *Ptosis operations.* See Chapter III.

§184. *Trichiasis and entropium operations.*—Among the many methods recommended for the cure of trichiasis or entropium of the eyelids two of the best seem to be those that have been perfected by *Green* and *Hotz*. (See Chapter III).

In *Green's* operation the conjunctiva and tarsal tissues are cut through in a horizontal direction parallel to the lid-margin and somewhat removed from it on the cutaneous surface. (See Fig.

FIG. 133.—(After Green). Shows the incision through the conjunctiva and tarsal tissue on the inner side of the upper lid in Green's operation for entropium and trichiasis.

133). Then a narrow strip of skin is removed from the outer surface of the eyelid corresponding to the tarsal incision. (See Fig. 134). Then sutures are put in in the following manner: a needle armed with a thread is entered near the ciliary margin and is brought out on the outer surface of the eyelid, at the lower wound-lip, and after gliding along on the tarsal tissue, it is brought out again through muscle and skin.   The ends are then tied.   Three or more such sutures are usually required.

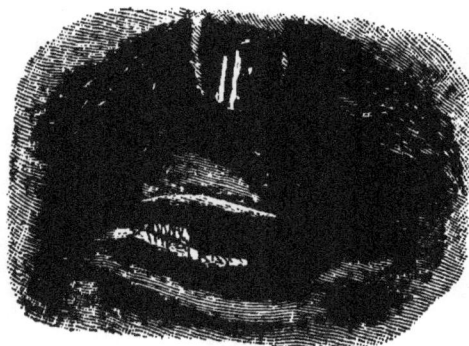

Fig 134.—(After Green). Shows the excision of a strip of skin from the upper lid in Green's operation for entropium and trichiasis.

In *Hotz's* operation an incision is made through the skin and muscle of the eyelid along the orbital edge of the tarsus, so as to lay bare the tarso-orbital aponeurosis.   Then a strip of muscular tissue is removed and sutures (three or four) are. applied in the following way: the needle, armed with a thread, is entered at the lower wound-lip through the skin and aponeurosis, and is brought out again through aponeurosis and skin at the upper wound-lip.   The ends are then tied so that the skin and aponeurosis'may heal together.

In order to do away with any loss of substance by the operative interference, it has been recommended (*von Millingen*) to. slit the lid-margin in two along its border so as to form two flaps, the outer consisting of the skin and cilia, the inner of tarsus and conjunctiva, and to implant a flap of skin or mucous membrane, or a *Thiersch* flap (*Gifford*) into this gap.

§185.   *Ectropium operations.*—Slight ectropium of the lower eyelid may be cured by *Adams'* operation (See Chapter III), or by removing a triangular piece from the tissue adjoining the outer angle of the palpebral fissure (See Fig. 135), so that the apex of the triangle lies somewhat higher than this angle.   Into the apex of the resulting gap the lower eyelid is

then drawn and stitched. *Kuhnt* recently recommended to remove a triangular piece of the conjunctiva and tarsal tissue,
not including the skin, its base lying at the ciliary margin ef
the eyelid, its apex in the fornix of the conjunctiva. This incision is followed by a single suture through the whole lid near
the ciliary margin.

FIG. 135.—Excision of a triangular piece of tissue from the outer canthus for the
cure of ectropium.

When ectropium is caused by the contraction of scar-tissue
after a cut or a deep burn, the excision of this scar-tissue may
sometimes be sufficient to cure the ectropium. In most cases,
however, it requires a more extensive operation, and usually
the transplantation of a flap or flaps, with or without a pedicle.

§186. *Canthotomy and canthoplasty.*—Shortening of the
palpebral fissure sometimes necessitates a surgical interference.
If the required effect need not be very large, canthotomy is
made. This consists in cutting through the tissues forming
the outer commissure of the eyelids by means of a strong pair
of scissors. To increase the effect of this little operation the
adjoining conjunctiva is undermined, and then stitched into
the gap resulting from the cut (canthoplasty). *Noyes* has advised a more effective mode of canthoplasty. The canthotomy being made, the section through the outer canthus is carried further on towards the temple. Then a small flap of skin
with its pedicle at the end of this cut is dissected and twisted,
so as to fit into the gap caused by the incision.

A description of the various methods for partial or total
blepharoplasty may be found in any modern text-book on operative surgery.

CHAPTER XXVII.—ON SPECTACLES.

§187. The best material for spectacle glasses is crown-glass, which is made by the fusion of white sand 120 parts, carbonate of potassium 35 parts, carbonate of sodium 26 parts, slaked lime or chalk 20 parts, and arsenic 1 part. *Flint glass* is heavier and softer. It consists of sand 42.5 parts, oxide of lead 43.5 parts, carbonate of potassium 11.7 parts, nitrate of potassium 1.8 parts, and chalk 0.5 parts (*Bohne*). Another material from which spectacle lenses are made is rock-crystal, commonly called pebbles. ·This is harder than glass, and consequently does not get scratched as easily.

Lenses are made by grinding a block of glass on tools which are segments of a sphere, or a cylinder, or a torus, to the desired curvature.

We have, accordingly, *convex spherical, concave spherical, convex cylindrical, concave cylindrical,* or combinations of these, and *toric* lenses. When a lens is convex (or concave) on one side and flat on the other it is called *plano-convex* or *plano-concave*; when it is convex (or concave) on both sides, it is called *bi-convex* (or *bi-concave*). In another form of lenses called *meniscus* or *periscopic* lenses, one side is convex and the other concave (*concavo-convex lens*). When one surface is plane and the other cylindrical, we speak of a *plano-cylindrical lens*; when one side is spherical and the other cylindrical, it is called a *sphero-cylindrical lens*. *Toric* lenses have two crossed cylinders of unequal radius ground on one side, while the other side may be plane or spherical (convex or concave). They are obtained by grinding the glass by means of a wheel the periphery of which has a curved (convex or concave) surface.

Another form of glasses which are sometimes made use of in spectacles alone or in combination with spherical or cylindrical surfaces, are *prismatic* glasses.

§188. Prisms refract the rays passing through them toward

their base, in consequence objects seen through them are apparently displaced from their real position. Spherical lenses may be considered as combinations of two prisms; in the bi-convex glass these prisms lie together with their bases, in the bi-concave lens they lie together with their refracting angles. Thus a bi-convex lens, refracting the rays towards the joint bases, renders them more convergent, while a bi-concave lens renders the rays passing through it less convergent or actually divergent, according to the larger or smaller distance of their source from the lens. The refracting action of cylindrical surfaces is the same, but only in the one meridian which stands at right angles to their axis.

§189. From the foregoing the action and value of different forms of spectacle lenses may be readily understood.

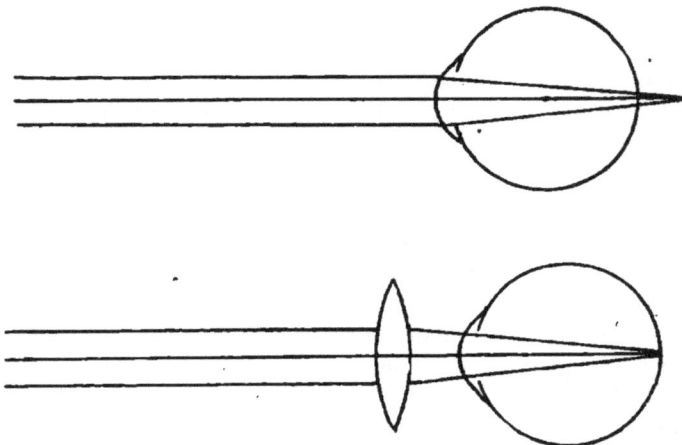

FIG. 136.—The upper figure shows the manner in which parallel rays are focussed by a hypermetropic eye behind its retina; the lower one shows the manner in which a convex lens alters the course of parallel rays so that they are focussed on the retina.

According to what was stated in Chapter XX, a hypermetropic eye is too short to be able to focus parallel rays on the retina (See Fig. 136), and in consequence its retina lies in front of the focus for parallel rays. A glass held in front of the eyes which will render the parallel rays more convergent, so as to bring their focus into the plane of the retina, will render this hypermetropic eye practically emmetropic. This is done by means of a convex lens.

In the same way, a concave lens will render a myopic eye practically emmetropic. A myopic eye is too long to be able to focus parallel rays on its retina. (See Fig. 137). Its retina, therefore, lies behind the focus for parallel rays. A glass, which, held before the eye, will render parallel rays so divergent that by the refraction of the eye their focus is just brought into the plane of the retina of the myopic eye, will correct the myopia. This must be a concave glass.

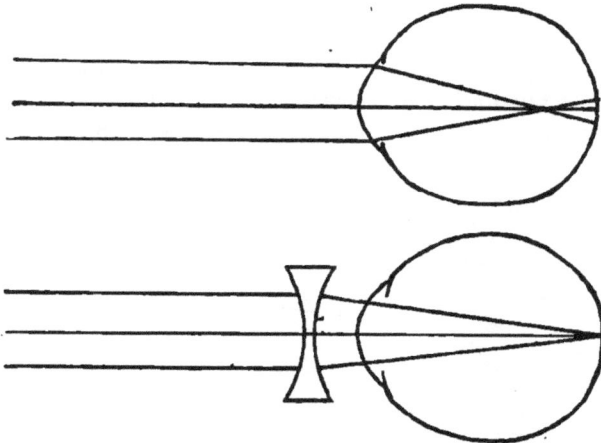

FIG. 137.—The upper figure shows how parallel rays are focassed by a myopic eye in front of its retina; the lower figure shows, how a concave lens alters the course of parallel rays, so as to be focussed on the retina of the myopic eye.

Cylindrical lenses have their place in the correction of astigmatism. As has been detailed, an astigmatic eye has a cornea which is curved asymmetrically. In the regular forms of astigmatism we have two so-called principal meridians which practically stand at right angles to each other, and one of which is the most while the other is the least curved one.

To render such an eye emmetropic we either must add to the refraction of the meridian of least curvature or reduce the refractive power of the meridian of highest curvature. Glasses which correct in one meridian only are cylindrical lenses. A cylindrical lens of the proper refracting power placed before the eye with its axis at right angles to the position of the meridian, the refraction of which is to be corrected, must bring about the desired result.

If an eye is built too short or too long, and has, moreover, an asymmetrically curved cornea, a combination of a cylindrical surface on the one side with a spherical one on the other side has to be made.

In presbyopia, the power of accommodation for near objects is wanting on account of the inability to make the crystalline lens convex enough through the action of the ciliary muscle. This means that the eye cannot render by its inherent faculty divergent rays, coming from near objects, convergent enough to bring their focus in the plane of the retina. This faculty can be equaled to a convex lens of such power as to focus divergent rays on the retina. (See Fig. 138). Conse-

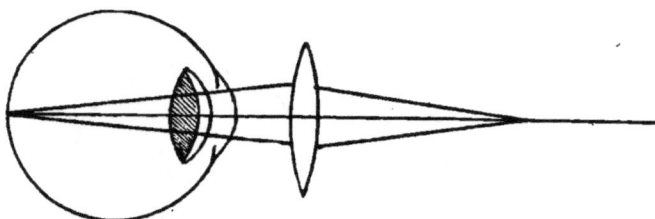

FIG. 138.—The (light) meniscus added to the (dark) crystalline lens represents the increase in its convexity during the act of accommodation. In presbyopia the crystalline lens cannot assume the convexity necessary for seeing near objects clearly. This lack of accommodation can be supplemented by putting a convex lens before the eye (which takes the place of the meniscus added to the crystalline lens when it can accommodate).

quently such a convex lens held before the eye will remedy presbyopia and enable the eye to see near objects plainly. The correcting lens for presbyopia in ametropic eyes of whatever degree and form must therefore, for near work, be added to the glass worn for distant vision.

In order to do away with the necessity of changing the glasses according to the near or far use they are put to, several kinds of glasses have been devised in which the two lenses are combined in one frame, so that the upper portion is the glass for the distant and the lower one the glass for near vision (two half lenses, *Franklin* glasses, the presbyopic lens pasted on the far lens or ground into it).

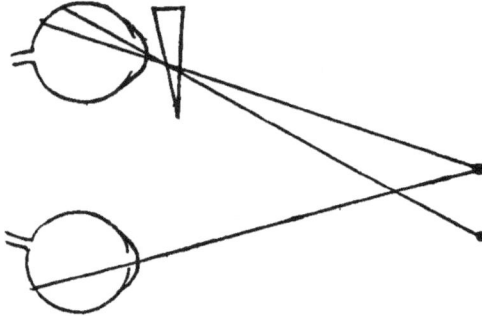

FIG. 139.—A prism placed before one eye with the base towards the temple will displace its image in such a manner as to cause heteronymous diplopia. The right image belongs to the left eye and the left image to the right eye. This is the manner in which a prism may correct diplopia due to divergence.

§190. Prismatic glasses, by their power of displacing the object looked at, may be made use of to correct diplopia. (See Fig. 139). A prism held before one eye with the base outward will cause crossed (heteronymous) diplopia; a prism held before one eye with the base inward will cause homonymus diplopia. (See Fig. 140). This is due to the fact that we project impressions received by the brain through the retina outward in the same direction in which they have come to the brain. An object looked at through a prism appears, therefore, to be displaced in the direction of the prolongation of the refracted rays. When looking through a prism this means that we see the object displaced toward the refracting angle or apex of the prism. We can, therefore, reduce double vision, caused by paresis or paralysis of a muscle, to single vision by placing a prism before the affected eye in such a manner as to neutralize the apparent displacement of objects looked at.

In the same manner we may in the case of insufficiency of a muscle supplement its deficient action by means of a prism, and, by not fully correcting it, help to strengthen such a muscle.

In conical cornea *hyperbolic* lenses, introduced by *Raehlmann*, have sometimes been found very valuable.

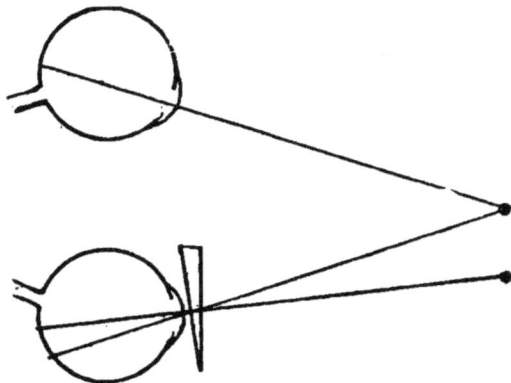

FIG. 140.—A prism with the base towards the nose held before one eye will displace its image in such a manner as to cause homonymous diplopia. The right image belongs to the right eye and the left image to the left eye. This is the manner in which a prism may correct diplopia due to convergence.

§191. Spectacles may be used also to simply protect an eye from the irritating rays of light, or from foreign substances apt to strike the eye deleteriously.

The former indication may be met by the use of smoke or blue glasses. Plain large-sized glasses are preferable to the curved ones which are often used. The curved glasses have but very seldom (*Eaton*) parallel surfaces, but are more often concave or concave cylindrical and, therefore, may be decidedly damaging. The coquilles with wire-netting should not be used at all, especially not in order to protect an inflamed eye, as they retain too much heat and prevent ventilation.

Workmen whose eyes are exposed to flying chips of metal or stone should wear protective spectacles made of mica, which are far superior to the wire-shields they sometimes use, since they allow of almost accurate vision and are tough enough to withstand a considerable amount of force.

CHAPTER XXVIII.—THE DRUGS MOST COMMONLY USED IN OPHTHALMIC PRACTICE.

§192. In order to produce mydriasis and paralysis of the accommodation, the chief drug is the sulphate of atropia. The physician, in prescribing it, ought to be sure that his druggist actually has the neutral salt, sulphate of atropia, in stock, and does not—as too often happens—simply dissolve atropia by means of sulphuric acid. A solution of atropia made with sulphuric acid is almost never perfectly neutral, and when the acid is in excess its instillation causes severe pain. To detect its acidity it may sometimes be necessary to allow the litmus-paper to remain in the solution for several hours.

℞ Atropiæ sulphatis neutr.,   -   grs. ij to iv.
    4% Sol. acid boracic.,   -   -   -   -   ℥j.
Sig.   To be dropped into the eye.

It will depend on the aim of the physician in the case in hand, how strong the solution must be, and how often it will have to be instilled. When complete paralysis of the ciliary muscle is desired for any reason, and in cases of iritis the strongest solution, grs. iv to ℥j, should be used, and the number and interval between the instillations is to be regulated according to necessity of the case, or in iritis to the severity of the inflammation, and the firmness of adhesions which may have been formed between the iris and the anterior lens-capsule.

For milder cases, for the examination of the back-ground of the eye, or for determining an error of refraction, etc., the less poisonous and milder mydriatic, hydrobromate of homatropine, has come into use. The form in which it is prescribed is the following:

℞ Homatropin. hydrobromat.,   -   grs. ij to iv.
    4% Sol. acid boracic,   -   -   -   -   ℥ij.
Sig.   To be dropped into the eye.

Both mydriatics may be used in the form of gelatine disks, or in solution in castor oil, or in the form of an ointment, and it is claimed that the effect of the mydriatic is greater and the quantity needed consequently smaller when so applied.

When the atropine is not well borne by the patient and causes symptoms of poisoning, it has long been recommended to use in its stead the extract of belladonna.

℞   Extract. Belladonn.,   -   -   -   grs. v.
     Aq. destillat.,   -   -   -   -   ℥ss.
Filter well.

Hyoscyamine, daturine and duboisine, which have been recommended as substitutes for atropine, seem to be in reality practically identical with this alkaloid.

A more effectual mydriatic than atropine is found in hyoscine, which, however, causes often a general intoxication.

℞   Hyoscin. hydroiod.,   -   -   -   grs. ij.
     4% Sol. acid. boracic.,   -   -   -   ℥ss.

Among the miotics (remedies which contract the pupil) eserine holds the first place.

℞   Eserin. sulph. neutr., or eserin salicyl., grs. j to ij.
     Aq. destillat.,   -   -   -   -   ℥ss.

Solutions of eserine when standing for some time, especially when exposed to light, turn brown-red. Such solutions, however, act as well and are not spoiled. A weaker effect may be reached by a $\frac{1}{2}$% solution of the muriate of pilocarpine.

Local anæsthesia is produced by cocaine (or tropa-cocaine). This drug may also with advantage be added to the atropine solution in iritis and painful keratitis, and by its mild mydriatic action is often sufficient for the examination of the fundus of the eye.

℞   Cocain. hydrochlor.,   -   -   -   grs. v.
     4% Sol. acid. boracic.,   -   -   -   ℥ij.

§193. For producing local antisepsis or a certain degree of asepsis, we use *bichloride of mercury, chlorinated water, boracic acid, pyoktanine, iodoform, aristol,* and other drugs.

℞ Hydrarg. bichlor.,   -   -   -   gr. $^1/_2$.
     Aq. destillat.,   -   -   -   -   ℨv.
Sig. To be poured into the eye three or four times a day.

℞ Aq. chlor.,   -   -   -   -   -   ℨj.
D. In a dark bottle.
Sig. To be dropped into the eye.

℞ Acid. boracic.,   -   -   -   -   -   ℨj.
     Aq. destillat.,   -   -   -   -   ℨiij.
Sig. To be poured into the eye every two or three hours.

℞ Pyoktan.,   -   -   -   -   -   gr. j.
     Aq. destillat.,   -   -   -   -   ℨj.
Sig. To be dropped into the eye.

This solution is very useful, if disagreeable, in affections of the lachrymal drainage apparatus.

Iodoform may be used in the form of an impalpable powder strewed into the eye, or in the shape of an ointment.

℞ Iodoform.,   -   -   -   -   grs. x to xx.
     Vasel. alb.,   -   -   -   -   ℨss.
Sig. To be rubbed into the eye once a day.

Aristol may also be used as a powder or in the form of an ointment.

℞ Aristol.,   -   -   -   -   -   grs. xx.
     Vasel. alb.,   -   -   -   -   ℨiij.
M. Sig. To be rubbed into the eye once a day.

§194. When an astringent action is desired *sulphate of zinc* answers the purpose best for milder action. If a stronger and superficial caustic action is desired, *nitrate of silver* is in its place. While the application of the zinc solution may, if desired, be safely left to the patient, the physician should apply the nitrate of silver always himself.

℞ Zinc. sulph.,　　-　　-　　-　　-　　grs. v.
　Aq. destillat.,　　-　　-　　-　　-　　ʒj.

To be brushed on the inside of the lower lids every morning.

℞ Argent. nitr.,　　-　　-　　-　　-　　grs. v.
　Aq. destillat.,　-　　-　　-　　-　　-　　ʒj.

Sig. To be brushed on the inside of the lids once a day.

In the treatment of trachoma *sulphate of copper* is the sovereign remedy (after squeezing out the granules). It is best used in the form of the pure crystal. Nicely shaped, smooth and round sticks can now be obtained ready-made at the druggist's.

§195. *Ointments*, of which various ones may be made use of in the treatment of external eye diseases, are best made with *white vaseline*. The common yellow vaseline is often acid and irritating.

℞ Hydrarg. ox. flav.,　　-　　-　　-　　grs. ij to iv.
　Vasel. alb.,　　-　　-　　-　　-　　-　　-　　ʒiij.

M. Sig. To be rubbed into the eyes.

This ointment of *yellow oxide of mercury*, which is very frequently used in ophthalmic practice, is better than a similar one made of the red oxide of mercury, as the yellow oxide is an impalpable powder. It must be so well mixed that no little grains or, worse yet, lumps of the powder can be seen in it. An ointment which is not well mixed is apt to cause a great deal of undue irritation and even ulceration.

Instead of white vaseline some prefer *lanoline* (wool-fat). My experiences with it have led me to discard its use years ago.

§196. There are three more remedies which are applied in the form of a powder. *Calomel (hydrarg. chlor. mite), iodoform* and *Jequirity*. Their use has been fully detailed in the chapters on the affections of the cornea and conjunctiva.

When it is desirable to create a vigorous diaphoresis, *extract of jaborandi* may be given internally. It is, however, better generally to use the alkaloid *pilocarpine* derived from it, in the form of the muriate of pilocarpine, for subcutaneous injections. It is always well first to give the patient a stimulant, and then inject the solution of pilocarpine muriate. I usually begin with gr. $\frac{1}{10}$ and increase the dose every day until disagreeable symptoms, as nausea and vomiting appear. This happens with most otherwise healthy people when gr. $\frac{1}{4}$ or $\frac{1}{3}$ has been injected, while others cannot stand more than gr. $\frac{1}{7}$ or $\frac{1}{6}$.

As adjuvants to the medicinal treatment in eye-affections we make use of heat, cold, natural and artificial leeches, massage, and the galvanic current.

INDEX.